Asphalt Pavements

Asphalt Pavements

Editor

Devendra Damale

Asphalt Pavements

Edited by **Devendra Damale**

Printed in 2017

ISBN: 978-1-68117-013-8

Library of Congress Control Number: 2015931397

Contents

vi

Preface

Asphalt pavements are easy to maintain, quick to construct, and provide a safe, smooth, quiet ride. Simply put, asphalt pavements provide the greatest level of drivability at the most economical price. Asphalt pavement refers to any paved road surfaced with asphalt. Hot Mix Asphalt (HMA) is a combination of approximately 95% stone, sand, or gravel bound together by asphalt cement, a product of crude oil. Asphalt cement is heated aggregate, combined, and mixed with the aggregate at an HMA facility. Asphalt is America's most recycled material. Reclaimed asphalt is not just reusable as a "black rock" – the asphalt cement in the reclaimed pavement is reactivated to become an integral part of the new pavement. The recycled asphalt cement replaces part of the new asphalt cement required for the pavement, reducing costs for road agencies.Recycling is just one reason that asphalt is the most sustainable pavement. Asphalt pavements that are designed and constructed as Perpetual Pavements never need to be removed and replaced. They are permanent structures. The only maintenance needed is infrequent replacement of the surface – and the material that is removed is recycled. This book focuses on asphalt pavement management, processes, operations, methods and techniques. Asphalt recycling also has been pointed out.

Editor

Road Asphalt Pavements Analyzed by Airborne Thermal Remote Sensing: Preliminary Results of the Venice Highway

Simone Pascucci[1,*], Cristiana Bassani[2], Angelo Palombo[1], Maurizio Poscolieri[3] and Rosa Cavalli[2]

[1]National Research Council, Institute of Methodologies for Environmental Analysis, C.da S. Loja -Zona Industriale, Tito Scalo (PZ), 85050, Italy

[2]National Research Council, Institute of Atmospheric Pollution, Via Fosso del Cavaliere, 100, Roma, 00133, Italy

[3]National Research Council, IDAC, Via Fosso del Cavaliere, 100, Roma, 00133, Italy

ABSTRACT

This paper describes a fast procedure for evaluating asphalt pavement surface defects using airborne emissivity data. To develop this procedure, we used airborne multispectral emissivity data covering an urban test area close to Venice (Italy).For this study, we first identify and select the roads' asphalt pavements on Multispectral Infrared Visible Imaging Spectrometer (MIVIS) imagery using a segmentation procedure. Next, since in asphalt pavements the surface defects are strictly related to the decrease of oily components that cause an increase of the abundance of surfacing limestone, the diagnostic absorption emissivity peak at 11.2μm of the limestone was used for retrieving from MIVIS emissivity data the areas exhibiting defects on asphalt pavements surface. The results showed that MIVIS emissivity allows establishing a threshold that points out those asphalt road sites on which a check for a maintenance intervention is required. Therefore, this technique can supply local government authorities an efficient, rapid and repeatable road mapping procedure providing the location of the asphalt pavements to be checked.

INTRODUCTION

According to the European Asphalt Pavement Association (EAPA; http://www.eapa.org/), asphalt pavement is commonly referred to as a mixture of bitumen and mineral matter [12,48] that for Italy asphalts is mainly composed of silicates and limestone [2]. The primary surface defects that occur on the asphalt pavement mixture are raveling, flushing and polishing [67]. Raveling is defined as "the progressive loss of pavement material from the asphalt surface caused by (a) stripping of the bituminous film from the aggregate, (b) asphalt hardening due to aging, (c) poor compaction especially or insufficient asphalt content". Flushing is the "excess asphalt on the surface caused by a poor initial asphalt mix design". Polishing is defined as "a smooth oily surface caused by traffic wearing off sharp edges of aggregates" [67].

The management and maintenance of transportation infrastructures are based on detailed and accurate information about the road network. The pavement type and road surface conditions are the most common variables required to provide detailed road mapping. This data is critical to the management decision process that involves billions of euros of assets, and maintenance budgets of millions of euros each year. Moreover, street maintenance work looks still today like something hurriedly thrown and only based on the roadman job. Often the maintenance is carried out when the pavement is approaching its collapse point and with renewal interventions linked to the worker experience [57].

Remote sensing can solve the road condition mapping by applying relatively cheap methods to evaluate the surface defects of asphalt pavements [25,29,63].

Haas et al. [19] were the first authors that investigated and extracted a pavement condition index (PCI) by connecting the road physical parameters (cracking, rutting and raveling) gathered from field observations with the Global Positioning System (GPS). This common technology provides detailed and geo-referenced information about road condition even though the low cost and easy managing requirements remain unsolved. Furthermore, land-based mobile mapping systems with an extensive set of sensors (including laser reflectometers, ultrasonic sensors, accelerometers, global positioning systems, gyroscopes, video and machine vision systems) and computers, such as Automated Road Analyser (developed by Roadware GRP of Paris, Ontario, Canada), have been commercially available for road mapping application [44,62]. However, the high cost of road network inspection requires the development of innovative remote sensing data analyses that are reasonably priced, easy to manage for the local authorities and valuable for the road condition mapping. For this purpose, many studies have been conducted to discriminate road surface distresses [25] and to analyze the spectral features of urban materials and their separability [4,20,22-24,51,56,58].

The characteristic absorption bands of silicates and limestone that outcrop when the asphalt pavements show surface defects, have been studied by many authors in the 8 to 12μm thermal infrared spectral region (TIR) [32-34,53-55,65,66].

Much progress has been made in understanding the nominal range of wavelengths suitable for detecting a variety of minerals and the relationship between spectral absorption feature intensity and mineral abundance using remote sensed data [27,32,34,40,41]. Among others, the TIR spectral region is becoming increasingly more important, thus enhancing the use of multi-channel remote sensing TIR instruments to discriminate geologic surface materials including carbonates, sulfates, clays, and felsic *vs.* mafic silicate minerals [3,14,49].

In this framework, the research will be focused on developing, implementing and validating the effectiveness of emission spectroscopy, in the TIR spectral range from 8.18μm to 12.70μm, to provide a rapid assessment of the asphalt surface distress. Fast and non-destructive methods, such as emission spectroscopy, offer potentially useful alternatives to time-consuming chemical methods of asphalt analysis. The characteristics of asphalt pavement emissivity spectra are controlled by mineral composition, water (hydration, hygroscopic, and free pore water) and particle size distribution.

Nowadays, the most common sensors that operate in the TIR range are: the TIMS instrument (8.2-12.6μm with 6 bands) [33], the SEBASS airborne sensor (7.57-13.5μm with 128 bands) [36-64], the ARES instrument (8.32-12.97μm with 32 bands) [52] and the AHI airborne sensor (7.5-11.7μm with 256 bands) [7]. Such spectral range is covered by sensors functioning also in other spectrum regions: the DAIS-7915, the Multispectral Infrared Visible Image Spectrometer (MIVIS), the AHS-160 and the MASTER simulator (0.46-2.39μm with 25 bands; 3.14-5.26μm with 15 bands; 7.76-12.87μm with 10 bands) [28].

In particular, the airborne MIVIS sensor, with its high spectral and/or spatial resolution [5], allows reliable quantitative measurements of specific absorption features of urban materials [1]. Moreover, a spatial resolution at least of 5m is optimum for urban applications [59,69], since "the spectral mixing space" becomes more complex with larger pixels [58,59]. Therefore, MIVIS sensor was used in order to retrieve a threshold to individuate those asphalt pavements to be checked for maintenance as it records emitted radiation in the TIR range, using a total of 10 bands with 2mrad of IFOV (Instantaneous Field Of View).

Since the asphalt pavements aging can be related to the loss of oily components [60] and to the sealing tar surface [25], and the decrease of oily components leads to an increase of several types of limestone deposits that are identifiable in the TIR range [36], a simple and fast method was developed in order to define a threshold on the basis of the band depth analysis [9] at 11.2μm (i.e., the limestone absorption peak in the TIR range).

For this purpose, an airborne MIVIS imagery covering a test area close to Venice city (Italy) was used for identifying in the TIR spectrum region the diagnostic asphalt emissivity features.

STUDY AREA

The study area (Figure 1) is characterized by a mixture of urban land cover types and surface materials, including many asphalted roads and, in particular, two main highways with asphalt pavements of different ages and conditions (i.e., more or less weathered and corroded). The study area corresponds to a MIVIS (Table 1) scene of 755 columns × 2956 lines (Figure 1b) and is centered at latitude 45°33′19″N and longitude 12°16′49″E. The flight strip was acquired over a rural area close to Venice city (Italy; Figure 1a) on November 23, 2006 at 11:56 (GMT), using scan rates of 25 scans/s at an altitude of 1500m, corresponding to a 3-m ground-pixel resolution at the instrument's IFOV.

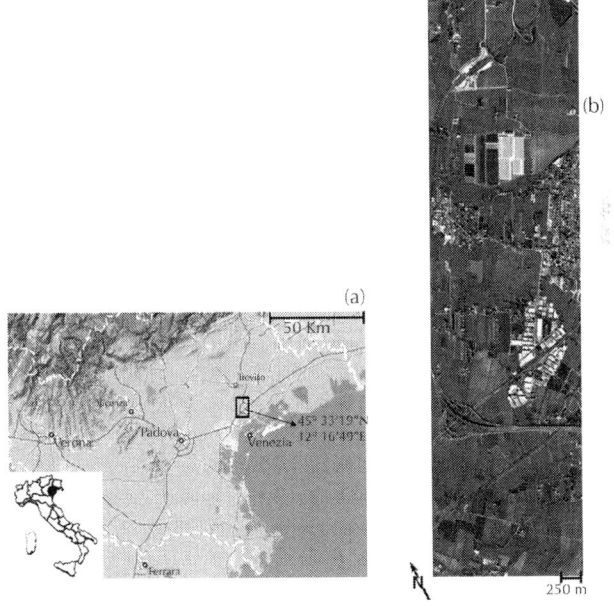

Figure 1: (a) MIVIS scene, outlined in black over a regional map; (b) MIVIS imagery acquired over Venice study area (755 columns × 2956 lines).

Table 1: MIVIS sensor characteristics.

	VIS: 0.43-0.83 µm (channels 1-20)		20 nm		6 - 366
Spectral coverage	NIR: 1.15-1.55 µm (channels 21-28)		50 nm		80 - 1062
	SWIR: 1.98-2.47 µm (channels 29-92)	Bandwidth	8 nm	SNR (min, max)	4 - 191
	TIR: 8.18-12.70 µm (channels 93-102)		340-540 nm		150 - 1500
FOV and IFOV	71° and 2 mrad	Cross-track pixels			755
Angular	1.64	Digitalization accuracy			12 bit

DATA AND METHODS

Image Preprocessing

MIVIS data preprocessing was performed as follows: (i) radiometric calibration of the raw data; (ii) atmospheric correction of the Thermal Infra-Red (TIR) data [18]; (iii) calibration to apparent emissivity by separating temperature and emissivity according to the methods described by [16,17].

The radiometric calibration of the airborne MIVIS TIR raw data was performed using a two-point calibration technique that is based on the linearity of the detector response over the dynamic range of the instrument [47]. To this goal, the maximum and minimum reference values of the radiance were acquired at the beginning and end of each scansion line to satisfy the calibration accuracy requirements.

The retrieved pixel spectral radiance (nW cm^{-2} sr^{-1} nm^{-1}) (R_λ^{pixel}), expressed by a linear equationobtained jointing the two reference points (i.e., the maximum and minimum reference values expressed by the radiance value and the corresponding Digital Number), is shown in Equation 1.

$$R_\lambda^{pixel} = (B_{\lambda,HBB} - B_{\lambda CBB}) \left[\frac{DN_\lambda^{pixel} - DN_{\lambda CBB}}{DN_{\lambda,HBB} - DN_{\lambda CBB}} \right] - B_{\lambda CBB} \qquad (1)$$

Where, the DNpixelλ is the pixel raw data to be radiometrically calibrated; the $DN_{\lambda CBB}$ and $DN_{\lambda HBB}$ are the spectral raw data measured for the cold (minimum reference value) and hot (maximum reference value) blackbody, respectively; $B_{\lambda CBB}$ and $B_{\lambda HBB}$ are the spectral radiance values for each blackbody as predicted by Planck's law with emissivity equal to one ($\varepsilon_\lambda = 1$) and known temperature (T_i).

As regards the atmospheric attenuation of the TIR spectral radiance that includes atmospheric transmission and upwelling atmospheric radiance, the ISAC (In-Scene Atmospheric Compensation) algorithm [31,36,64,70] was employed for MIVIS TIR atmospheric correction. This algorithm assumes two pixels of the scene to be blackbodies on which neither locations nor temperatures are known. The ISAC algorithm is suitable also when the atmospheric radiative conditions are not available during the acquisition time. For this study, we followed the "most hits" method as described by Johnson [31] and pixels whose emissivity was equal to 1 at the wavelength were used as a marker.

Once the image was atmospherically corrected, we used the method developed by Johnson [31], and revised by Hook et al. [27] and Kahle et al. [34] to retrieve apparent emissivities. The method is based on the Planck's law for gray body radiator ($\varepsilon_\lambda \neq 1$). The standard formulation of this law for the spectral radiance (L_λ) of each pixel (i) is described by the following equation:

$$L_{\lambda,i} = [\varepsilon_{\lambda,i} B_\lambda(T_i) + (1 - \varepsilon_{\lambda,i})L_{SKY_\lambda}]\tau_\lambda + L_{ATM_\lambda} \qquad (2)$$

where $\varepsilon_{\lambda,i}$ is the surface spectral emissivity of pixel (i); $B_\lambda(T_i)$ is the blackbody spectral radiance at Ti temperature, located at pixel (i); $L_{SKY\lambda}$ is the spectral downwelling radiance; τ_λ is the spectral atmospheric transmission; $L_{ATM\ \lambda}$ the spectral upwelling radiance. The $L_{SKY\ \lambda'}$ and $L_{ATM\ \lambda}$ are related to the emission of the atmosphere itself that reaches directly the sensor or is reflected by the surface before being acquired by the sensor and they are also assumed to be independent from the view angle (i.e., pixel location) [36,64].

In order to extract pixel emissivity from the pixel radiance, the requirement is to discriminate from Equation (2) the emission that depends on the kinetic T and other atmospheric parameters as well as the ε value for each image band [16]. This multivariable problem is solved by using the assumption of ISAC algorithm about the linear relationship between the observed radiance and the Planck's function, whose slope is related to the atmospheric transmission (τ_λ), and whose offset is the upwelling atmospheric radiance $(L_{ATM\,\lambda})$, at that wavelength:

$$B_\lambda(T_i) = \frac{[L_{\lambda,i} - L_{ATM_\lambda}]}{\tau_\lambda \varepsilon_{\lambda,i}} \tag{3}$$

The emissivity can be calculated from this linear equation, in terms of the other variables:

$$\varepsilon_{\lambda,i} = \frac{[L_{\lambda,i} - L_{ATM_\lambda}]}{[B_\lambda(T_i)\tau_\lambda]} \tag{4}$$

Substituting in Eq. (4) the Eq. (2) and solving for T, then gives

$$T_i = \frac{C_2}{\lambda \ln[(\varepsilon_{\lambda,i} C_{1\tau\lambda} / \pi\lambda^5 (L_{\lambda,i} - L_{ATM_\lambda})) + 1]} \tag{5}$$

As the goal of this study is the assessment of the spectral feature shapes and the band depth analysis, we need to retrieve relative emissivities. In this context, several methods have been proposed in literature for deriving emissivities such as the reference channel method [32], the emissivity normalization method [17], the temperature-independent spectral indices [3], the thermal log residuals and alpha residuals [26], and the spectral emissivity ratios [68]. Several of these methods are compared and reviewed by Gillespie [15-17], by Hook et al. [26,27] and by Li et al. [45].

In this study we applied the emissivity normalization routine proposed by Realmuto [50], Hook et al. [26], Kealy and Hook [35] and Gu et al. [18] and implemented in the ENVI 4.4. [41] image processing software. This routine, first, derives the brightness temperature of each pixel from the pixel radiance. Afterward, the apparent emissivity image is obtained by normalizing the radiance of each pixel to the Planck's curve that is generated from the pixel with the maximum brightness temperature with an emissivity value set to 0.96 (i.e., a reasonable hypothesis for exposed mineral surfaces).

Image Classification

To develop a method for automated image analysis of road asphalt pavements on the basis of their emissivity spectral features, (a) an object-oriented approach and (b) a band-depth analysis were used (Figure 2).

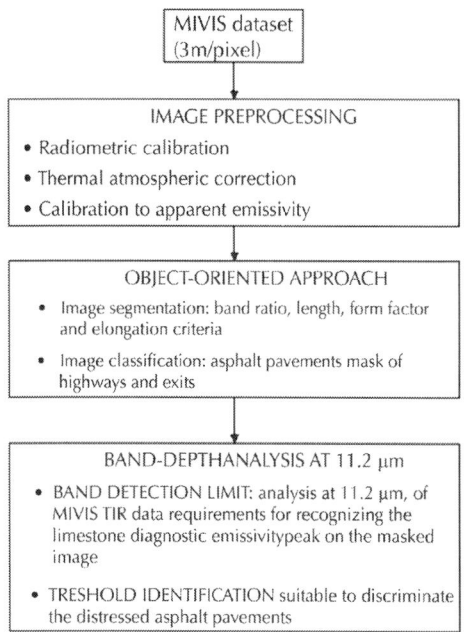

Figure 2: Flow diagram indicating the steps followed in the methods.

Object-Oriented Approach

An object-oriented approach was first adopted in order to individuate and select on MIVIS imagery the asphalt roads [6,43], which was also used as an input mask later on in the band-depth analysis.

Object-oriented approaches can represent a valuable alternative to the conventional pixel-based classification methods, as they consider the spatial context. Segmentation methods divide a study area into adjoining clusters of pixels, called segments or regions, based on similarity or dissimilarity of their single or multiple-layer pixel values [61].

The segmentation approach allows for: (a) quantifying the spatial heterogeneity within the data at various scale levels; (b) delineating homogeneous patches also involving a certain spatial generalization; (c) implementing an explicit hierarchal structure between segments at different spatial scales. As a result, spatial information is very important in classification processes to produce reliable maps [46].

For this study, according to Jensen [30], we used an object-oriented approach with a segmentation procedure followed by classification as implemented in the Feature Extraction module of the ENVI 4.4 software package [42]. In more detail, the procedure consists of a combined process of segmenting the image into regions of pixels, computing attributes for each region to create objects, and last classifying the objects. In order to identify only road asphalt pavements with the requirement of a sufficient neighboring number of pixels showing a homogeneous asphalt mixture, we chose only highways and exits asphalt pavements for the further band-depth analysis. For this purpose, a workflow consisting of two main tasks was adopted.

(*i*)   The *"find objects"* task (i.e. segmentation; [30]) that was divided, in its turn, into four steps: "segment", "merge", "refine", and "compute attributes". The "segment" and "merge" steps of this task were used to divide the image into segments corresponding to real-world objects and for solving over-segmentation problems and then the adjacent segments were grouped on the basis of their brightness value.

(*ii*) The *"rule-based classification"* task (i.e. classification; [30]) was used to extract only the highways and exits objects and then to export them onto a raster image.

In this final step, the rules criteria with the relative appropriate scale factors (i.e., weights) were identified using both spectral and spatial attributes for classifying the highways/exits asphalt pavements objects.

Band Depth Analysis on Asphalt Roads

In Italy road paving asphalts are made of a mixture of mineral aggregate (mainly of silicates and limestone granules) and bitumen [2]. Asphalt pavements surface defects [67] are strictly related to the decrease of the oily components of the bitumen, thus increasing the surface abundance of the limestone granules (see Figure 3). Therefore, for this study we chose the outcrop of the limestone granules, which certainly highlights distressed paving asphalts, as an indicator for those asphalt pavements to be checked for maintenance.

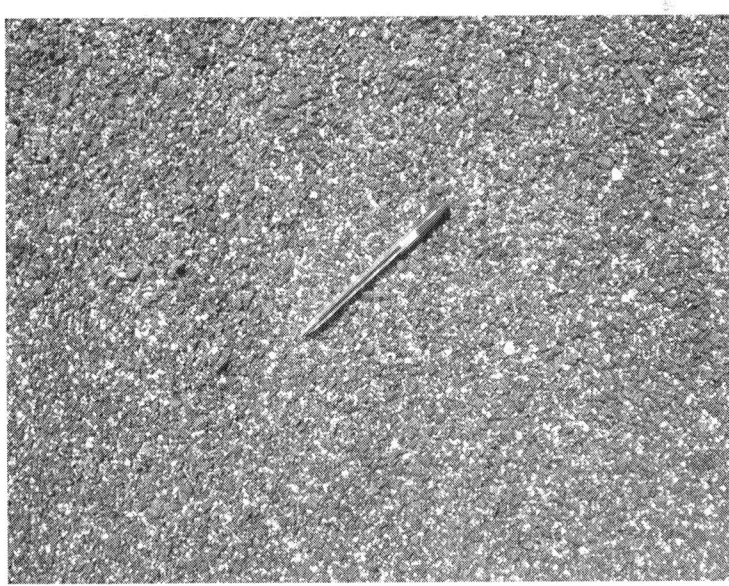

Figure 3: Example of an asphalt pavement of the study area with surfacing limestone granules.

The applied procedure consists, first, of the analysis of the available John Hopkins University (JHU) spectral library [55] information allowed for retrieving the emissivity spectral features (8-13μm TIR range) of the asphalt paving material. An absorption band can be described by characteristics, such as the position, depth, width and asymmetry [8-10,68]. The presence of an absorption feature and its position in the reflectance/emissivity spectrum provides valuable information about the chemical composition of a material [21].

Figure 4 depicts the JHU emissivity spectra convolved to MIVIS bandpasses in order to show how their occurrence would affect MIVIS detectability of the major limestone absorption feature. Looking at Figure 4a, it is evident that the main difference between the emissivity spectral feature, in the 8-13µm range, of new and old pavement asphalts is the spectral contrast centered at 11.2µm. Moreover, the study of Kirkland et al. [36] confirms that the 11.2µm is the diagnostic emissivity band for the limestone (Figure 4b).

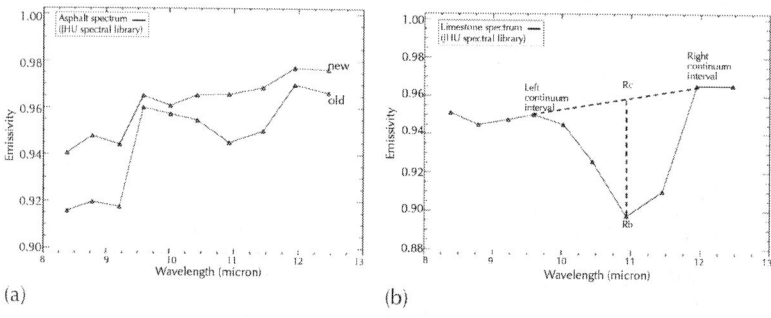

(a)   (b)

Figure 4: Examples (a) of new and old emissivity spectra of paving asphalt from the JHU spectral library and (b) of limestone band-depth analysis (intervals 9.59-11.94µm): emissivity continuum-removed absorption peak of a pure limestone spectrum (JHU spectral library), both convolved to MIVIS bandpasses in order to show how its occurrence would affect MIVIS detectability.

In view of this information, a methodology was developed to map by airborne remote sensing the asphalt roads showing surface defects on the basis of the diagnostic 11.2µm limestone emissivity peak.

Therefore, once selected the highway/exits asphalt roads by means of the object-oriented approach, a Band-Depth (BD) analysis was performed on MIVIS emissivity data to measure the spectral contrast of the limestone peak at 11.2µm that is centered, if convolved to MIVIS bandpasses, at 10.93µm (Figure 4b).

The BD analysis primarily requires application of the continuum-removal method that consists of: (a) "fitting a straight line hull to represent the reflectance or emissivity background using two continuum tie points on either side of the absorption feature" [8,39] and (b) dividing the spectrum by this fitted continuum line. Thus, continuum-removed absorption features can be directly overlapped to one another by scaling them to the same depth at the band centre allowing a comparison of the shapes of their absorption features.

The absorption band depth (D) was calculated from:

$$D = 1 - Rb / Rc \qquad (6)$$

where Rc is the emissivity of the continuum at the band centre and Rb is the emissivity at the band centre (Figure 4b).

RESULTS AND DISCUSSION

Object-oriented Classification Results

In this application, the highways/exits asphalt roads feature classification, shown in Figure 5, was obtained by assigning to the band ratio criterion (concerning the MIVIS emissivity channels centered at 10.93 and 11.94μm) a weight of 0.5, whereas the spatial criterion received the remaining weight of 0.5 (length 0.4, form factor 0.2 and elongation 0.4). These optimal segmentation parameters were determined using a systematic trial and error approach validated by the visual inspection of the quality of the output image objects, i.e., how well the asphalt pavements matched feature boundaries in the image [46]. Moreover, on the obtained classification image some residual errors were manually corrected on the image in order to achieve a mask of only asphalted highways and exits (Figure 5).

250 m

Figure 5: Object-oriented classification of MIVIS emissivity image. In yellow are depicted the masked highways and exits, they are overlaid on MIVIS channel 13 only for visualization purposes.

This object-oriented approach used for the extraction of asphalt roads is very cost-effective, because it reduces the necessity for laborious on-screen digitizing that is by far the most expensive task of the standard photo-interpretation process.

Application Requirements and Band-Depth Results

In order to be confident that MIVIS instrument technical characteristics allow to recognize the peculiar asphalt absorption features at 11.2μm (i.e., the outcropping limestone granules) and to apply BD analysis, a Band Detection Limit (BDL) was first calculated according to the Kirkland et al. [36,38] method.

BDL is the percentage of the absorption features, in our case the 11.2μm limestone peak, to be detected with the desired confidence level [38]. The resulting value depends on the BDL related to the width and depth of the spectral contrast exhibited by the target. The BDL is calculated using:

$$BDL = \frac{CF}{SNR\sqrt{\dfrac{BandFWHM}{SamplingInterval}}} \tag{7}$$

where BDL = the minimum band depth required for detection of a given band width and center; Confidence Factor (CF) = the contrast relative to the SNR level that a feature should exhibiting to be distinguished from background; Band FWHM = the full-width target band at the half maximum of the band depth; Sampling Interval = the instrumental spectral sampling interval.

According to Kirkland et al. [38], the BDL value depends on the instrument signal-to-noise ratio (SNR), instrument spectral resolution, target spectral band depth, band width, desired signal level above the noise (CF) and atmospheric compensation. Lower numbers for the BDL indicate that lower spectral contrast is required for detection. The CF influences the BDL such that a higher CF requires greater band contrast for acceptance (i.e., a CF = 1 represents a signal level comparable with noise).

The BDL values were calculated by assuming that the target surface covered the instrument's field of view, and no atmospheric attenuation influenced the data. It is clear, however, that even though the noise levels for a given sensor are generally fixed, for remote sensing data application, the signal portion of the SNR is affected by other external factors such as view angle, atmospheric attenuation and scattering and surface emissivity, which can modify the sensor perceiving signal [11].

Since the sensor SNR [13] cannot be modeled without specific knowledge of the instrument characteristics, the following calculations were made using MIVIS sensor characteristics to establish its capability to retrieve spectral absorption features in the TIR region.

The MIVIS sensor DN values, acquired in-flight on the internal blackbodies, were converted to emissivity following the same procedure used in § 3.1 [26,35], except for atmospheric correction. Atmospheric correction of the blackbody data was not taken into account, because the distance between the sensors and the blackbodies was negligible. The SNR of the TIR bands was, consequently, estimated by dividing the emissivity mean spectra of all masked asphalt pavement pixels (i.e., signal) by the standard deviation of blackbody emissivity (i.e., noise). Figure 6shows MIVIS SNR calculated for the TIR spectral range on the masked asphalt pavements of the study area.

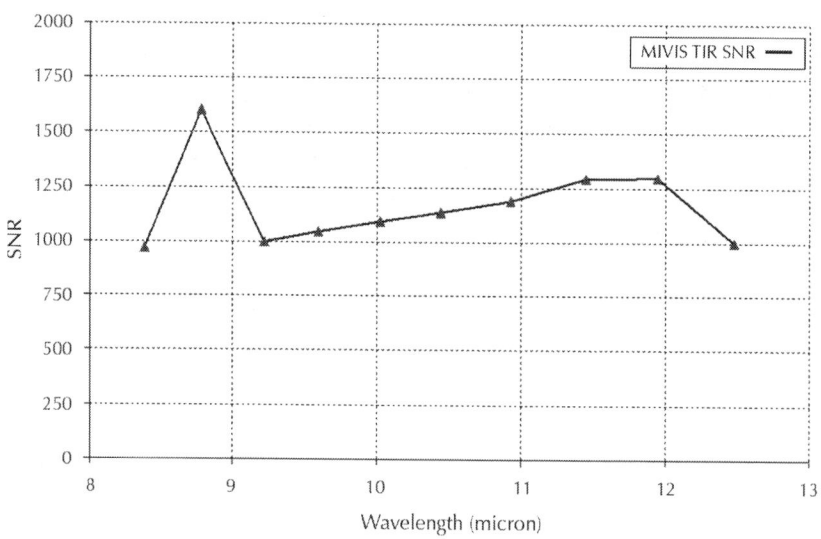

Figure 6: Estimates of MIVIS SNR in the TIR spectral range calculated on the masked asphalt pavements of the study area.

Regarding the spectra recorded by the MIVIS scanner on the masked asphalt pavements, to have a spectral feature that can be detected with the desired CF, the asphalt material has to depict a spectral contrast greater than the MIVIS scanner BDL. Therefore, a BDL value of 0.0012 was obtained by using, in Equation 7, MIVIS instrument characteristics within the desired TIR interval (as shown in Figure 4b), i.e., a FWHM of 0.4 μm, a band sampling interval at 10.93 □ μm of 0.2μm, a SNR at 10.93 um above 1100 and with a CF of 2.

The 0.0012 BDL value is the minimum limestone fractional exposure on the asphalt material required for being detected by MIVIS instrument characteristics.

Once the BDL analysis confirms that MIVIS characteristics allows for recognizing the limestone diagnostic emissivity peak, the MIVIS emissivity data of the masked highways/exits pavements were analyzed by means of the band-depth procedure to establish a suitable threshold level for discriminating those asphalt pavements to be checked for maintenance intervene.

For this purpose, an extensive field survey was carried out on all masked asphalt roads to visually check the asphalt pavements conditions. Figure 7a shows the two selected areas with certainly surface defects on which retrieving the BD threshold level that identifies those asphalt pavements (i.e., pixels) where to check for an asphalt maintenance intervention. On the two selected test areas the distribution function of the BD (at 10.93μm) analysis was calculated. Next, a threshold, which is based on the BD values distribution and that allows for discriminating in both test areas distressed asphalt pavements pixels (visually checked by field surveys), was assessed by using:

(a)   (b)

Figure 7: (a) In yellow are depicted the two test areas, selected for training the band-depth analysis; (b) Image showing the band-depth analysis results: in red are depicted the detected asphalt pavements showing surface defects thus to be checked for maintenance. Both images are overlaid on MIVIS channel 13 only for visualization purposes.

$$\mu - \sigma \qquad\qquad\qquad (8)$$

where, μ is the mean value of the test areas BD distribution and σ is the corresponding standard deviation. As a consequence, the BD value of 0.020 is the identified threshold for determining on all masked pixels (i.e., the highways/exits) the asphalt pavements to be checked for maintenance work. Figure 7b shows in red the distressed asphalt pixels individuated using the above mentioned threshold of 0.020. For example, in Figure 8, asphalt pavements of a highway and the relative exits, different from the two chosen as test areas, are shown as a particular case as they demonstrate the ability of the proposed procedure in detecting the different surface defects of the same asphalt.

(a)   (b)

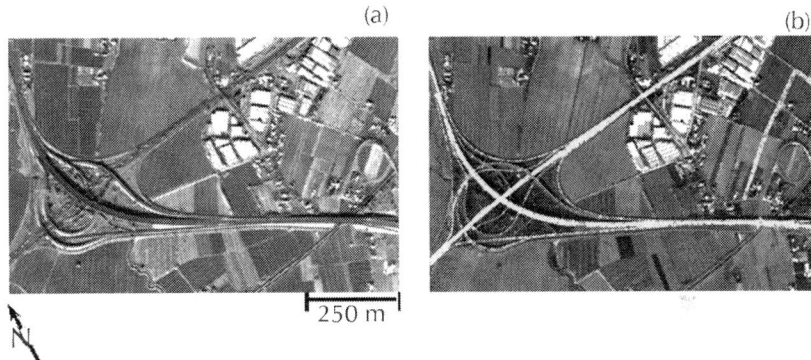

250 m

Figure 8: Images showing an example of asphalt pavements with different surface defects within the study area. Image (b) shows MIVIS emissivity BD classification results. Both images are overlaid on MIVIS channel 13 only for visualization purposes.

All the areas classified as distressed (depicted in red in Figure 7b) were further visually checked by means of a field survey carried out on the accessible roads mostly for safety reasons. Based on the visual field accuracy assessment, the asphalts to be checked for maintenance can be properly identified in all the masked study area with accuracy ranging from 75 to 90%, corresponding respectively to a total surface area of about 70,200 m² and 84,240 m² (total area is 93,600m²).

In particular, both MIVIS classification and field surveys allowed observing that surface defects were greater for the highways exit asphalts than for the highways' ones.

CONCLUSIONS

The preliminary results of this study demonstrate the utility of airborne remote sensing for identifying asphalt roads and discriminating those on which checking for a maintenance work. The potentiality of assessing the surface defects of asphalt pavements by using airborne Thermal Infra-Red data was established. This method can be surely applied to roads (e.g. highways, secondary roads) crossing rural environments and in large parking areas and wide streets within urban centres, whereas in case of densely settled urban areas it is difficult to successfully exploit this procedure. In fact, the presence of heat islands and other major drawbacks (e.g., brightness effects) would not allow obtaining suitable results.

The integration of object-oriented classification (i.e., segmentation followed by classification) applied to MIVIS multispectral TIR data permitted the identification of all the road asphalt pavements with a very cost-effective and accurate procedure.

For the surface defects assessment of the identified asphalt roads we chose the surfacing limestone granules occurrence as a suitable indicator. Therefore, on the primary emissivity absorption feature of the limestone at 11.2µm a band-depth analysis of the continuum-removed absorption feature was performed on a MIVIS airborne TIR imagery. The detection limit analysis showed that MIVIS technical characteristics in the TIR range allow discriminating this spectral feature. However, a more accurate discrimination of the limestone diagnostic absorption feature can be achieved using more advanced hyperspectral thermal instruments, such as AHI and SEBASS airborne sensors.

In conclusion, the proposed combination of segmentation procedure and emissivity shape-based analysis used in this preliminary study allows for a rapid discrimination (i.e., a suitable threshold) of the location of distressed asphalt pavements. Therefore, the encouraging results may let public institutions and private companies to adopt this procedure, applied to airborne remote sensing data, to rapidly control and monitoring the surface defects of the road asphalt pavements. Such large-scale monitoring can be also included in the more general framework of critical and civil infrastructures management and protection.

Future application of the proposed procedure will include field spectral analyses, using a portable micro-Fourier transform infrared radiometer: (a) to construct a spectral library of asphalt pavements with different surface defects useful for calibrating and validating further thermal remote sensing data acquired on this study area and on other test sites; and (b) to provide asphalt samples for laboratory mineral abundance analysis and for determining their deterioration/aging level.

We would like to thank an anonymous reviewer, whose thoughtful comments helped to improve the final manuscript.

REFERENCES

1. Bassani, C.; Cavalli, R.M.; Cavalcante, F.; Cuomo, V.; Palombo, A.; Pascucci, S.; Pignatti, S. Deterioration status of asbestos-cement roofing sheets assessed by analyzing hyperspectral data. *Remote Sensing of Environment* 2007, *109*, 361–378.

2. Bassi, P. *Chimica applicata ai materiali da costruzione* 1993.

3. Becker, F.; Li, Z.L. Temperature-Independent Spectral Indices in Thermal Infrared Bands. *Remote Sensing of Environment* 1990, *32*, 17–33.

4. Ben-Dor, E.; Levin, N.; Saaroni, H. A spectral based recognition of the urban environment using the visible and near-infrared spectral region (0.4–1.1 m). A case study over Tel-Aviv. *International Journal of Remote Sensing* 2001, *22*(11), 2193–2218.

5. Bianchi, R.; Marino, C. M.; Pignatti, S. Airborne hyperspectral remote sensing in Italy. Proceedings of Recent Advances in Remote Sensing and Hyperspectral Remote Sensing, Rome, Italy, September 23-30; 1994; pp. 29–37.

6.   Boskovitz, V.; Guterman, H. An adaptive neuro-fuzzy system for automatic image segmentation and edge detection. *IEEE Transactions on fuzzy systems* 2002, *10*, 247–262.

7.   Carlisle, O.; Lucey, P.G.; Sherman, S.B. Thermal infrared weathering trajectories in Hawaiian basalts: results from airborne, field and laboratory observations. Proceedings of the 37th Lunar and Planetary Science Conference, League City, Texas, March 13-17; 2006; 17.

8.   Clark, R.N.; Roush, T.D. Reflectance Spectroscopy: Quantitative Analysis Techniques for Remote Sensing Applications. *Journal of Geophysical Research* 1984, *89*, 6329–6340.

9.   Clark, R.N.; Gallagher, A.J.; Swayze, G.A. Material absorption band depth mapping of imaging spectrometer data using the complete band shape least-squares algorithm simultaneously fit to multiple spectral features from multiple materials. Proceedings of the 3th JPL Airborne Visible/Infrared Imaging Spectrometer (AVIRIS) Workshop: JPL Publication, Pasadena, CA,, June 4-5; 1990; 90-54, pp. 176–186.

10.   Clark, R.N. Spectroscopy of rocks and minerals and principles of spectroscopy. In *Manual of Remote Sensing. Volume 3: Remote Sensing for the Earth Sciences*; Rencz, A.N., Ed.; John Wiley & Sons Inc., 1999; pp. 3–58.

11.   Colwell, R.N. *Manual of Remote Sensing*; American Society of Photogrammetry and Remote Sensing: Falls Church Eds., 1983; Volume 1196, pp. 344–363.

12.   CONCAWE. Bitumens and bitumen derivatives. In *CONservation of Clean Air and Water in Europe (PD 92/104)*; Brussels, Belgium, 1992.

13.   Gao, B. An operational method for estimating signal to noise ratios from data acquired with imaging spectrometers. *Remote Sensing of Environment* 1993, *43*, 23–33.

14.   Gillespie, A.R.; Kahle, A.B.; Palluconi, F.D. Mapping alluvial fans in Death Valley, CA, using multispectral thermal infrared images. *Geophysical Research Letters* 1984, *11*(11), 1153–1156.

15.   Gillespie, A.R. Spectral mixture analysis of multispectral thermal infrared images. *Remote Sensing of Environment* 1992, *42*, 137–145.

16. Gillespie, A.R.; Rokugawa, S.; Matsunaga, T.; Cothern, J.S.; Hook, S.; Kahle, A.B. A Temperature and Emissivity Separation Algorithm for Advanced Spaceborne Thermal Emission and Reflection Radiometer (ASTER) Images. *IEEE Transactions on Geoscience and Remote Sensing* 1998, *36*(4), 1113–1126.

17. Gillespie, A.R.; Rokugawa, S.; Hook, S.; Matsunaga, T.; Kahle, A.B. ASTER Temperature/Emissivity Separation Algorithm Theoretical Basis (Version 2.4). In *Algorithm Theoretical Basis Document*.; Washington, DC: NASA Contract Number NAS5-31372. , 1999.

18. Gu, D.G.; Gillespie, A.R.; Kahle, A.B.; Palluconi, F.D. Autonomous Atmospheric Compensation (AAC) of high-resolution hyperspectral thermal infrared remote-sensing imagery. *IEEE Transactions Geoscience Remote Sensing* 2000, *38*(6), 2557–2570.

19. Haas, R.; Hudson, W. R.; Zaniewski, J. *Modern Pavement Management*; Krieger Publishing Company: Malabar, FL, 1994.

20. Heiden, U.; Roessner, S.; Segl, K.; Kaufmann, H. Analysis of spectral signatures of urban surfaces for their area-wide identification using hyperspectral HyMap data. Proceedings of IEEE -ISPRS Joint Workshop on Remote Sensing and Data Fusion over Urban Areas, Rome, Italy, November 8-9; 2001; pp. 173–177.

21. Heiden, U.; Segl, K.; Roessner, S.; Kaufmann, H. Determination of robust spectral features for identification of urban surface materials in hyperspectral remote sensing data. *Remote Sensing of Environment* 2007, *111*, 537–552.

22. Hepner, G.F.; Chen, J. Investigation of imaging spectroscopy for discriminating urban land covers and surface materials. Proceedings of AVIRIS Earth Science and Applications Workshop, Palo Alto, CA, 27 Feb - 2 Mar; 2001.

23. Herold, M.; Gardner, M.; Roberts, D. Spectral resolution requirements for mapping urban areas. *IEEE Transactions on Geoscience and Remote Sensing* 2003, *41*(9), 1907–1919.

24. Herold, M.; Roberts, D.A.; Gardner, M.E.; Dennison, P.E. Spectrometry for urban area remote sensing. Development and analysis of a spectral library from 350 to 2400 nm. *Remote Sensing of Environment* 2004, (91), 304–319.

25. Herold, M.; Roberts, D. Spectral characteristics of asphalt road aging and deterioration: implications for remote-sensing applications. *Applied Optics* 2005, *44*(20), 4327–4334.

26. Hook, S.J.; Gabell, A.R.; Green, A.A.; Kealy, P.S. A comparison of techniques for extracting emissivity information from thermal infrared data for geologic studies. *Remote Sensing of Environment* 1992, *42*, 123–135.

27. Hook, S. J.; Abbott, E. A.; Grove, C.; Kahle, A. B.; Palluconi, F. D. Use of multispectral thermal infrared data in geological studies. In *Manual of Remote Sensing. Volume 3: Remote Sensing for the Earth Sciences*; Rencz, A.N., Ed.; John Wiley & Sons Inc., 1999; pp. 59–110.

28. Hook, S. J.; Meyers, J. J.; Thome, K. J.; Fitzgerald, M.; Kahle, A. B. The MODIS/ASTER airborne simulator (MASTER) – a new instrument for earth science studies. *Remote Sensing of Environment* 2001, *76*(1), 93–102.

29. Jensen, J.R.; Cowen, D.C. Remote Sensing of Urban/Suburban Infrastructure and Socio-economic Attributes. *Photogrammetric Engineering and Remote Sensing* 1999, *65*(5), 611–622.

30. Jensen, J.R. *Introductory Digital Image Processing: A Remote Sensing Perspective*, 3rd Ed. ed.; Upper Saddle River, NJ: Prentice Hall, 2005; p. 526.

31. Johnson, B.R. scene atmospheric compensation: Application to SEBASS data collected at the ARM site, Part I. *Aerospace Corporation technical report, ATR-99 (8407)-1* 1998.

32. Kahle, A.B.; Rowan, L.C. Evaluation of multispectral middle infrared aircraft images for lithologic mapping in the East Tintic Mountains, Utah. *Geology* 1980, *8*, 234–239.

33. Kahle, A.B.; Goetz, A.F. Mineralogic information from a new airborne thermal infrared multispectral scanner. *Science* 1983, *222*, 24–27.

34. Kahle, A.B.; Palluconi, F.D.; Christensen, P.R. Thermal emission spectroscopy: application to Earth and Mars. In *Remote geochemical analysis: elemental and mineralogical composition*; Pieters, C.M., Englert, P.A.J., Eds.; Cambridge University Press, 1993; pp. 99–120.

35. Kealy, P.S.; Hook, S.J. Separating temperature and emissivity in thermal infrared multispectral scanner data: implications for recovery of land surface temperatures. *IEEE Transactions on Geoscience and Remote Sensing* 1993, *31*, 1155–1164.

36. Kirkland, L.E.; Kenneth, C.H.; Salisbury, J.W. Thermal Infrared spectral band detection limits for unidentified surface materials. *Applied Optics* 2001, *40*(27), 4852–4864.

37. Kirkland, L.; Kenneth, H.; Keim, E.; Adams, P.; Salisbury, J.; Hackwell, J.; Treiman, A. First use of an airborne thermal infrared hyperspectral scanner for compositional mapping. *Remote Sensing of Environment* 2002, *80*, 447–459.

38. Kirkland, L.E.; Herr, K.C.; Adams, P.M. Infrared stealthy surfaces: Why TES and THEMIS may miss some substantial mineral deposits on Mars and implications for remote sensing of planetary surfaces. *Journal of Geophysical Research* 2003, *108*(E12), 5137.

39. Kokaly, R.F.; Clark, R.N. Spectroscopic determination of leaf biochemistry using band-depth analysis of absorption features and stepwise multiple linear regression. *Remote Sensing of Environment* 1999, *67*, 267–287.

40. Kruse, F.A.; Boardman, J.W.; Huntington, J.F. Fifteen years of hyperspectral data: Northern Grapevine Mountains, Nevada. In *Proceedings of the 8th JPL Airborne Earth Science Workshop: JPL Publication*; Jet Propulsion Lab: Pasadena, CA, 1999; Volume 99-17, pp. 247–256.

41. Kruse, F.A.; Boardman, J.W.; Huntington, J.F. Comparison of Airborne Hyperspectral Data and EO-1 Hyperion for Mineral Mapping. *IEEE Transactions on Geoscience and Remote Sensing* 2003, *41*(6), 1388–1400.

42. ITT Visual Information Solutions. ENVI - Environment for Visualizing Images, Version 4.4. 2008.

43. Lhermitte, S.; Verbesselt, J.; Jonckheere, I.; Nackaerts, K.; van Aardt, J.A.N.; Verstraeten, W.W.; Coppin, P. Hierarchical image segmentation based on similarity of NDVI time series. *Remote Sensing of Environment* 2007, *112*, 506–521.

44. Li, J.; Chapman, M.A. *Terrestrial mobile mapping systems towards real-time geospatial data collection*; Geospatial Information Technology for Emergency Response. ISBN978-0-415-42247-5,

ISPRS Book Series; Taylor & Francis: London, 2008; Volume 6, pp. 103–119.

45.   Li, Z.L.; Becker, F.; Stoll, M. P.; Wan, Z. Evaluation of six methods for extracting relative emissivity spectra from thermal infrared images. *Remote Sensing of Environment* 1999, *69*, 197–214.

46.   Mathieu, R.; Aryal, J.; Chong, A.K. Object-Based Classification of Ikonos Imagery for Mapping Large-Scale Vegetation Communities in Urban Areas. *Sensors* 2007, *7*, 2860–2880.

47.   Palluconi, F. D.; Meeks, G.R. Thermal infrared Multispectral scanner (TIMS): an investigarot's guide to TIMS data. In *JPL Publication*; Jet Propulsion Laboratory: Pasadena, CA, 1985; Volume 85-32.

48.   Puzinauskas, V.P.; Corbett, L.W. Differences between petroleum asphalt, coal-tar pitch, and road tar. *Asphalt Institute (RR 78-1)* 1978.

49.   Ramsey, M.S.; Christensen, P.R. Mineral abundance determination: Quantitative deconvolution of thermal emission spectra. *Journal of Geophysical Research* 1998, *103*, 577–596.

50.   Realmuto, V.J. Separating the Effects of Temperature and Emissivity: Emissivity Spectrum Normalization. In *Proceedings 2th TIMS Workshop, JPL Publication*; Jet Propulsion Laboratory: Pasadena, CA, 1990; Volume 90-55, pp. 31–35.

51.   Roessner, S.; Segl, K.; Heiden, U.; Kaufmann, H. Automated differentiation of urban surfaces based on airborne hyperspectral imagery. *IEEE Transactions on Geoscience and Remote Sensing* 2001, *39*(7), 1525–1532.

52.   Richter, R.; Müller, A.; Habermeyer, M.; Dech, S.; Segl, K.; Kaufmann, H. Spectral and radiometric requirements for the airborne thermal imaging spectrometer ARES. *International Journal of Remote Sensing* 2005, *26*(15), 3149–3162.

53.   Ruff, S.W.; Christensen, P.R.; Barbera, P.W.; Anderson, D.L. Quantitative thermal emission spectroscopy of minerals: A laboratory technique for measurement and calibration. *Journal of Geophysical Research* 1997, *102*(14), 899–913.

54.   Sabine, C.; Realmuto, V.J.; Taranik, J.V. Quantitative estimation of granitoid composition from Thermal Infrared Multispectral Scanner (TIMS) data, desolation wilderness, Northern Sierra Nevada,

California. *Journal of Geophysical Research* 1994, *99*(B3), 4261–4271.

55. Salisbury, J.W.; Walter, L. S.; Vergo, N.; D'Aria, D.M. *Infrared Spectra of Minerals (2.1- 25 micrometers).*; Johns Hopkins University Press, 1991.

56. Segl, K.; Heiden, U.; Roessner, S.; Kaufmann, H. Fusion of spectral and shape features for identification of urban surface cover types using reflective and thermal hyperspectral data. *ISPRS Journal of Journal of Photogrammetry & Remote Sensing* 2003, *58*, 99–112.

57. SITEB. *Quaderno tecnico per la manutenzione delle pavimentazioni stradali.*; SitebSì, srl., Ed.; 2004.

58. Small, C. Scaling Properties of Urban Reflectance Spectra. Proceeding of AVIRIS Earth Science and Applications Workshop, Pasadena, CA, 27 Feb - 2 Mar; 2001.

59. Small, C. High spatial resolution spectral mixture analysis of urban reflectance. *Remote Sensing of Environment* 2003, *88*, 170–186.

60. Speight, J.G. Asphalt. In *Kirk-Othmer's Encyclopedia of Chemical Technology*; Kroschwitz, J.L., Howe-Grant, M., Eds.; John Wiley & Sons Inc., 1992; pp. 689–724.

61. Stuckens, J.; Coppin, P. R.; Bauer, M. E. Intergrating contextual information with per-pixel classifications for improved land cover classifications. *Remote Sensing of Environment* 2000, *71*, 282–296.

62. Tao, C.V.; Li, J. *Advances in Mobile Mapping Technology*; ISPRS book Series; Taylor & Frances: London ISBN 978-0-415-42723-4. , 2007; Vol. 4, p. 176.

63. Usher, J.M. Remote sensing applications in transportation modelling. *Remote Sensing Technologies Center Final Report*, 2000.

64. Vaughan, R.G.; Wendy, M.C.; Taranik, J.V. SEBASS hyperspectral thermal infrared data: surface emissivity measurement and mineral mapping. *Remote Sensing of Environment* 2003, *85*, 48–63.

65. Vincent, R. K.; Thomson, F. Spectral composition imaging of silicate rocks. *Journal of Geophysical Research* 1972, *77*, 2465–2472.

66.   Vincent, R.K.; Thomson, F.; Watson, K. Recognition of exposed quartz sand and sandstone by two-channel infrared imagery. *Journal of Geophysical Research* 1972, *77*, 2473–2477.

67.   Walker, D.; Entine, L.; Kummer, S. *Pavement surface evaluation and rating. Asphalt PASER manual.* Wisconsin Transportation Information Center, 2002.

68.   Watson, K. Spectral ratio method for measuring emissivity. *Remote Sensing of Environment* 1992, *42*, 113–116.

69.   Welch, R. Spatial resolution requirements for urban studies. *International Journal of Remote Sensing* 1982, *3*(2), 139–146.

70.   Young, S. J. In scene atmospheric compensation: Application to SEBASS data collected at the ARM Site. Part II. *Aerospace Corporation technical report, ATR-99 (8407)-II* 1998.

chapter 2

Performance of Recycled Asphalt Pavement as Coarse Aggregate in Concrete

Fidelis O. OKAFOR

Department of Civil Engineering, University of Nigeria, Nsukka, Nigeria

ABSTRACT

Recycled asphalt pavement (RAP) is the reclaimed and reprocessed pavement material containing asphalt and aggregate. Most RAP is recycled back into pavements, and as a result there is a general lack of data pertaining to the mechanical properties for RAP in other possible applications such as Portland cement concrete. In the present study, some mechanical properties of Portland cement concrete containing RAP as coarse aggregate were investigated in the laboratory. Six concrete mixes of widely differing water/cement ratios and mix proportions were made using RAP as coarse aggregate. The properties tested include the physical properties of the RAP aggregate, the compressive and flexural strengths of the concrete. These properties were compared with those of similar concretes made with natural gravel aggregate. Results of the tests suggest that the strength of concrete made from RAP is dependent on the bond strength of the "asphalt-mortar" (asphalt binder-sand-filler matrix) coatings on the aggregates and may not produce concrete with compressive strength above 25 MPa. However, for middle and low strength concrete, the material was found to compare favorably with natural gravel aggregate.

INTRODUCTION

Concrete is one of the most widely used construction material today. The concrete consists at least 75% by volume of aggregate materials which may be locally available but in some places it may be economical to substitute those natural aggregates by more cheaply and abundantly available materials. Several comprehensive studies over the years have dealt with the subject of aggregate supplies and needs and the possible use of waste materials as aggregates for concrete. Attempts have been made therefore to replace natural aggregates in conventional concrete by locally available materials such as sintered domestic refuse [1], palm kernel shell [2], palletized blast furnace slag [3] and most widely recycled concrete [4- 10]. Nevertheless, critical shortage of natural aggregate for production of concrete is still developing in many regions and the need for better methods of solid waste disposal and probably energy conservation have contributed to the increased interest in this technology. In most third world countries where technological development is still growing, some regions especially large urban areas already have a problem in obtaining adequate aggregate supplies at reasonable cost. At the same time, increasing quantities of demolished asphalt pavement materials from road reconstruction projects are generated as a waste material close to these areas. These waste asphalt pavement materials are usually plowed back as sub-base material during the reconstruction process or used as embankment fill material which does not represent the most suitable use for the RAP. One of the possible ways to enhance the ample use of RAP would be to incorporate the material into Portland cement. However, little research has been done [11, 12] to explore the potential of incorporating RAP into concrete. This paper is a discussion of the results of tests carried out to assess the performance of RAP as coarse aggregate in concrete. The performance in concrete of RAP was compared with that of uncrushed natural gravel aggregate.

MATERIALS AND PROCEDURES

Materials

Commercially available Type I Portland cement (Dangote brand) conforming to BS 12 [13] was used in this study and the specific gravity of the cement was 3.14. The fine aggregate used in all the tests was river sand. The grading of the sand conformed to the zone 3 requirements of BS 882 [14]. The natural coarse aggregate used in the investigation is uncrushed natural gravel. The physical properties of the coarse aggregate are shown in Table 1. The tests were carried out in accordance with BS 812 [15].

Preparation and Physical Properties of RAP Aggregate

RAP coarse aggregate was prepared by crushing waste asphalt pavement rubbles obtained from a waste dump at a road reconstruction site along Enugu-Onitsha expressway, Nigeria. The old asphalt pavement was a fine-graded, hot-mix asphaltic concrete. The waste asphalt pavement rubbles were crushed, sieved and graded to sizes similar to that of the natural gravel. The RAP coarse aggregate consist therefore of "asphalt-mortar" (asphalt binder-sand-filler matrix) coated aggregates retained on 4.75 mm sieve. The physical properties of RAP aggregate are shown in Table 1. The tests were carried out in accordance with BS 812 [15].

Preparation of Specimens

Two mix proportions of 1:2:4 and 1:3:6 by weight of cement, sand and RAP aggregate were made with water/cement ratios of 0.50, 0.60

and 0.70. In addition, control mixes were made with natural aggregate having identical proportion by weight of cement, sand, coarse aggregate and water/cement ratio as their respective mixes containing RAP. Workability test was carried out on the various concretes by slump test in accordance with BS 1881 [16] standard test methods. All test specimens were kept under cover with wet jute bags in the laboratory until demolding at 24 hours after which they were transferred to curing water at room temperature. The properties of the hardened concrete tested were the compressive strength on 100 mm cubes and the flexural strength on 100 mm × 100 mm × 500 mm beams with third-point loading, three samples being tested at each age of 7, 14, 28, 56 and 91 days.

RESULTS AND DISCUSSIONS

Aggregate

It can be seen from the results shown in Table 1 that compared to the natural aggregate, RAP has lower specific gravity and lower water absorption. The lower specific gravity of RAP may be attributed to the considerable amounts of low density asphalt-mortar (asphalt binder-sand-filler matrix) coatings on the recycled aggregate which reduced the overall density of the material (RAP). The slightly lower water absorption of RAP may also be attributed to the asphalt coating which prevented full absorption of water by the aggregate.

Table 1: Physical Properties of Coarse Aggregates

Physical properties	Gravel	RAP
Specific gravity	2.70	2.28
Water absorption, (%)	3.2	2.9
Aggregate crushing value, (%)	20.1	
Aggreate impact value, %	8.2	4.3
Grading (% by weight .passing: sieve stated)		

25.4 mm	100	100
20.0 mm	98	98
12.5 mm	88	85
9.6 mm	57	58
4.75 mm	1	0
2.36 mm	0	

The aggregate crushing value test performed on RAP gave no measurable result as the test specimen under loading, rather than fragmenting tend to compress into a single dense mass. The reason could be probably due to the presence of soft asphalt binder which under the confined pressure tend to flow round the individual aggregates and thus binding the compressed and probably crushed aggregate into a single solid mass. It is then evident from the result that the crushing value test prescribed in BS 812 [15] cannot be used to assess the physical strength of RAP aggregate, a more appropriate testing method is necessary. The aggregate impact value test result shown in Table 1 indicates a superior value for RAP compared to the natural gravel. It would be wrong however, to assume on the basis of the result that the physical strength of RAP is higher than that of the natural gravel. Rather the test result may be a probable indication that RAP is a less brittle material and can absorb more impact load than the natural aggregate. The higher impact resistance of RAP may be attributed to the presence of softer and more elastic asphalt-mortar (asphalt binder-sand-filler matrix) coatings on the recycled aggregate which probably inhibited crack propagation through the material. Thus, the developing crack rather than go through the aggregate is spread over the more elastic asphalt-mortar, during which more energy is absorbed. This may likely be the reason for the toughness improvement of concrete made with RAP [12].

Workability

The results of slump test for all concrete mixes are presented in Table 2. The results indicate that RAP concrete is less workable than natural aggregate concrete for the same free water/cement ratio. The lower workability of RAP concrete may be attributed to the high viscosity of asphalt-mortar coating on the aggregate. It is also believed that the lower workability of RAP concrete may be as a result of the rather rough irregular shape of the aggregate, the gravel aggregate having more rounded shape. Nevertheless, RAP concrete was very workable and could easily be mixed and compacted.

Table 2: workability test results

Mix proportion	Water/cement ratio	Slump, (mm)	
		Gravel	RAP
1:2:4	0.50	70	33
1:2:4	0.60	84	45
1:2:4	0.70	100	74
1:3:6	0.50	41	17
1:3:6	0.60	50	30
1:3:6	0.70	80	40

Compressive Strength

From the results of the development of compressive strength up to the age of 91 days presented in Figure 1 and 2 for all mixes, it is evident that in every case the strength continues to increase with age and that, for the given water/cement ratio and mix proportion, concretes made from the natural aggregate have a higher strength than those from RAP aggregate at all ages.

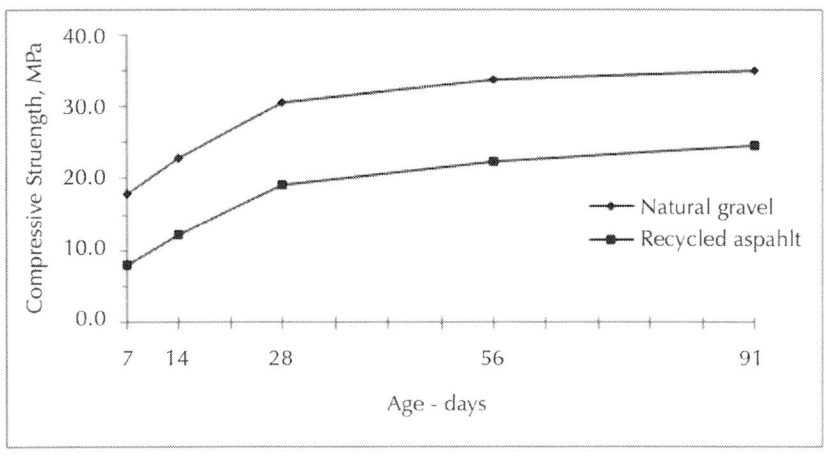

Figure 1.a: Water/cement ratio 0.50.

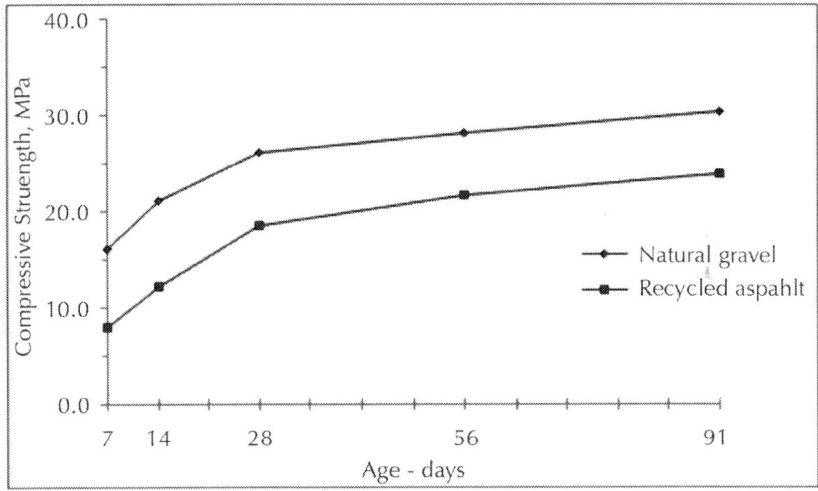

Figure 1.b: Water/cement ratio 0.60.

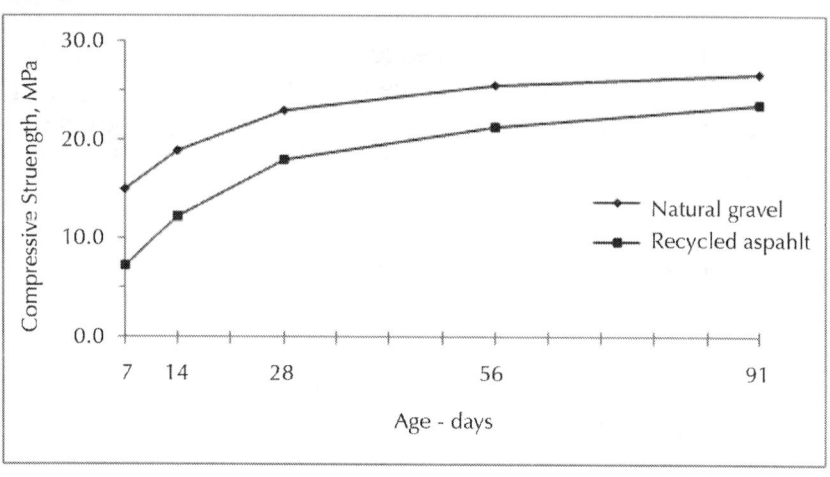

Figure 1.c: Water/cement ratio 0.70.

Figure 1: Relationship between compressive strength and age (mix 1:2:4).

A closer examination of Figure 1 reveals that the strength of RAP concrete remained almost unaltered despite the reduction of the water/cement ratio suggesting that no further improvement in strength above 24.6 MPa may be achieved by reducing the water/cement ratio below 0.50. This trend may be explained by arguing that the strength of these concretes is governed primarily by the strength of the RAP aggregate which in turn is dependent to a large extent on the bond strength of the asphalt-mortar (asphalt binder-sand-filler matrix) attached to the aggregate particles. Failure in compression is probably initiated by the apparent weak bond between the asphalt-mortar and aggregate. This fact was established by an examination of the failure surface of the broken cubes. With the concretes made from RAP aggregate, failure in compression was dominated by a breakdown in bond between the aggregates and the attached old asphalt-mortar without any apparent crushing of the aggregate while for the natural aggregate concretes of similar mixes crushing of aggregate was more prominent. The strength of the control concrete made from the natural aggregate at water/cement ratio of 0.50 was on average of 11 MPa higher than that made from RAP at all ages. The difference in strength is however, observed to decrease as the water/cement ratio increases and was on average of 4.5 MPa at water/cement ratio of 0.70. The reason for this is that the strength of the control mix alone has been reduced due to the increased water/cement ratio with the subsequent reduction in cement mortar and bond strengths. The performance of the two types of concrete is also compared in Figure 2 but the concretes in this case have been made leaner with significantly higher aggregate/cement ratio (1:3:6 by weight of cement, sand and coarse aggregate).

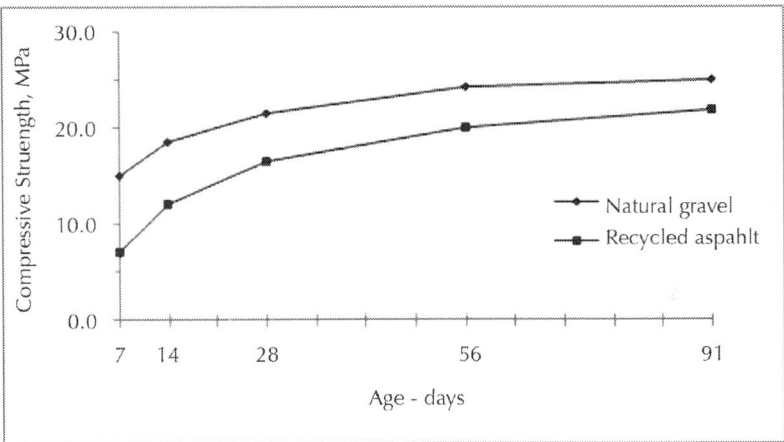

Figure 2.a: Water/cement ratio 0.50.

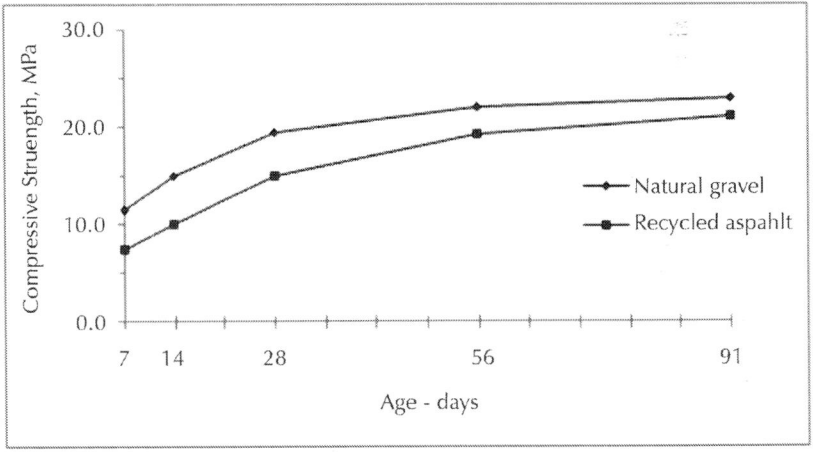

Figure 2.b: Water/cement ratio 0.60.

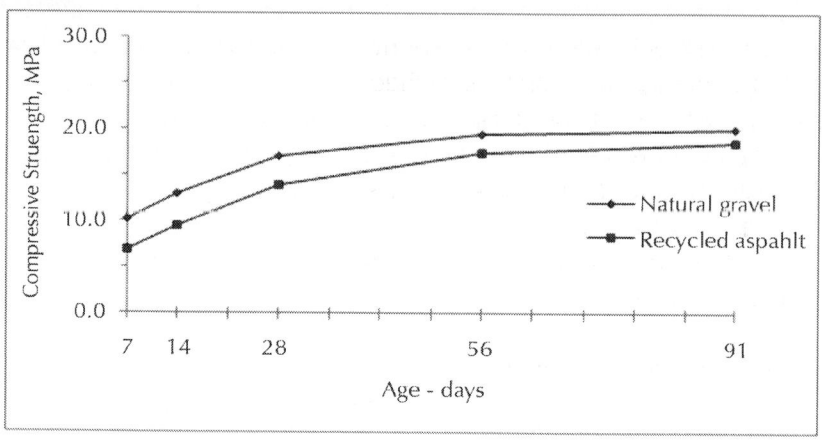

Figure 2.c: Water/cement ratio 0.70.

Figure 2: Relationship between compressive strength and age (mix 1:3:6).

An examination of Figure 2 shows that, unlike the previous case of the richer mix, both concretes show a reduction in strength with increase in water/cement ratio, the natural aggregate concrete exhibiting the greater reduction. It is evident then, that with such a lean mix, the strength of the cement mortar has been reduced below that of the asphalt mortar/aggregate bond. Compressive strength of RAP concrete is now dependent more on the strength of the cement mortar rather than on the strength of the aggregate just as with the control mix. Hence, the difference between the two becomes considerably reduced as the water/cement ratio increases and is on average of 4.2 MPa and 1.5 MPa for water/cement ratio of 0.50 and 0.70 respectively.

FLEXURAL STRENGTH

Results of the development of flexural strength up to the age of 91 days for all mixes are shown in Figure 3 and 4. The trends of the data are very similar to those of the compressive strength. As with the compressive strength, Figure 3 indicates that flexural strength is dependent more on the bond strength of the asphalt-mortar attached to the aggregate particles. The increase in water/cement ratio of the mix from 0.50 to

0.70 had very marginal effect on the flexural strength of RAP concrete since the flexural strength of these mixes is dependent on the bond strength of the asphalt-mortar attached to the aggregate which remain unchanged. In Fig. 4 the concrete is made leaner by increasing the aggregate/cement ratio and the bond strength of the asphalt-mortar becomes less dominant in the over-all failure mechanism, failure is now governed more by the strength of the cement mortar-aggregate bond, as with the control mix and hence their strengths are more closer together with increase in water/cement ratio.

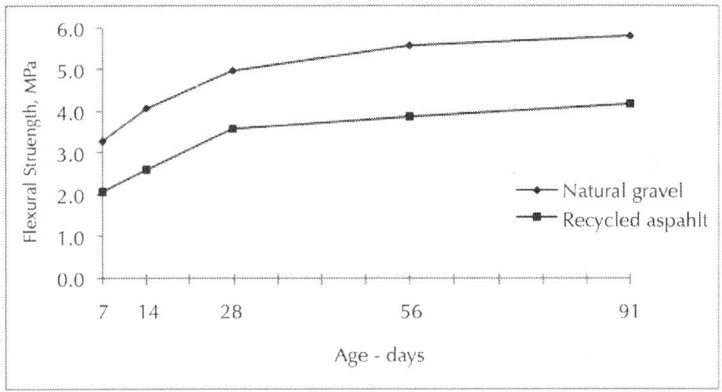

Figure 3.a: Water/cement ratio 0.50.

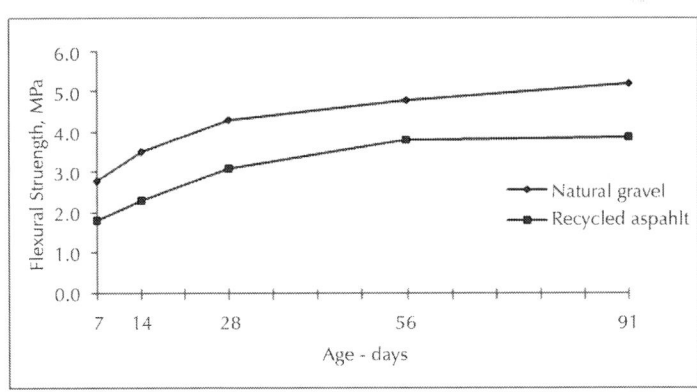

Figure 3.b: Water/cement ratio 0.60.

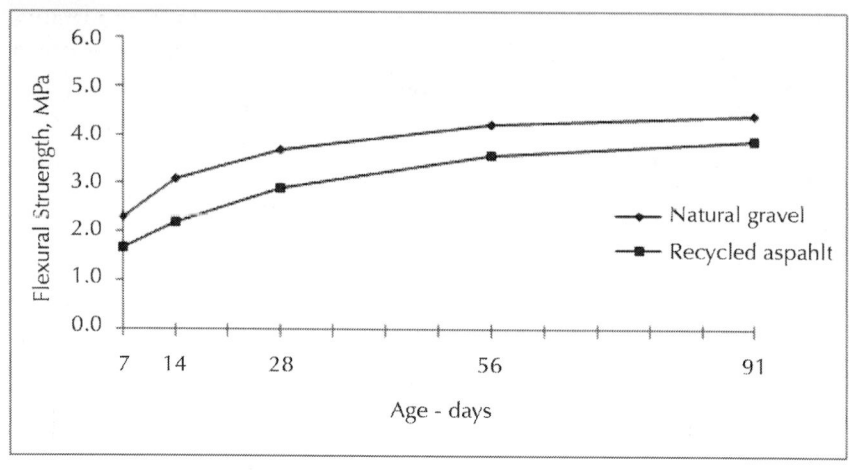

Figure 3.c: Water/cement ratio 0.70.

Figure 3: Relationship between flexural strength and age (mix 1:2:4).

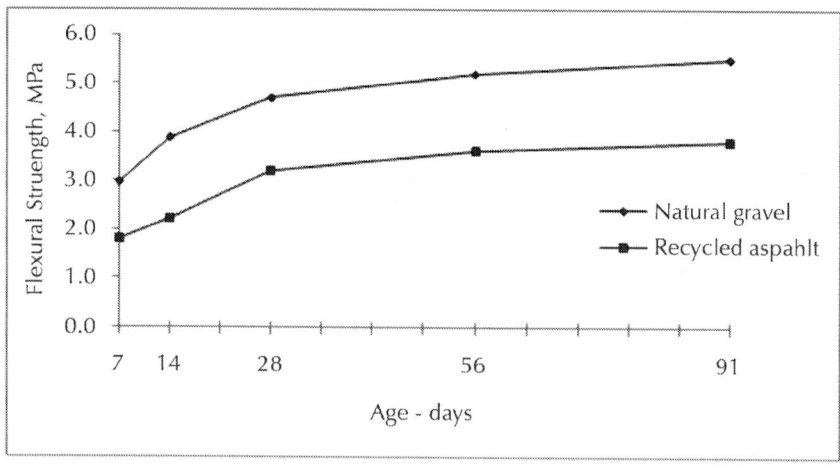

Figure 4.a: Water/cement ratio 0.50.

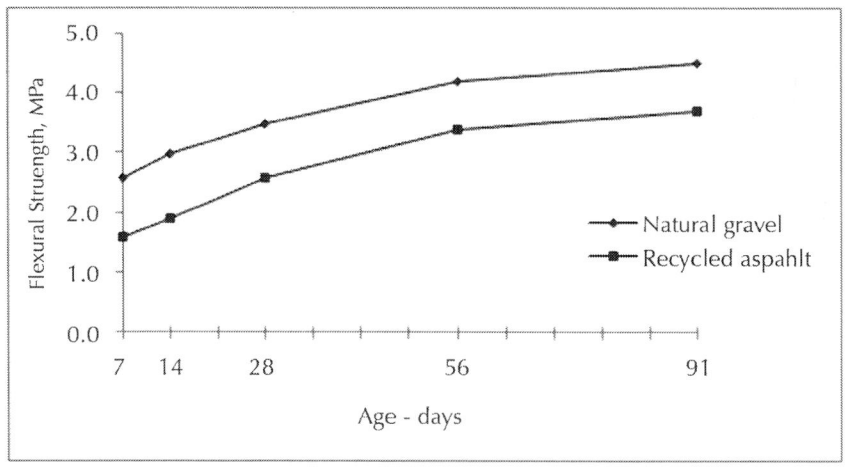

Figure 4.b: Water/cement ratio 0.60.

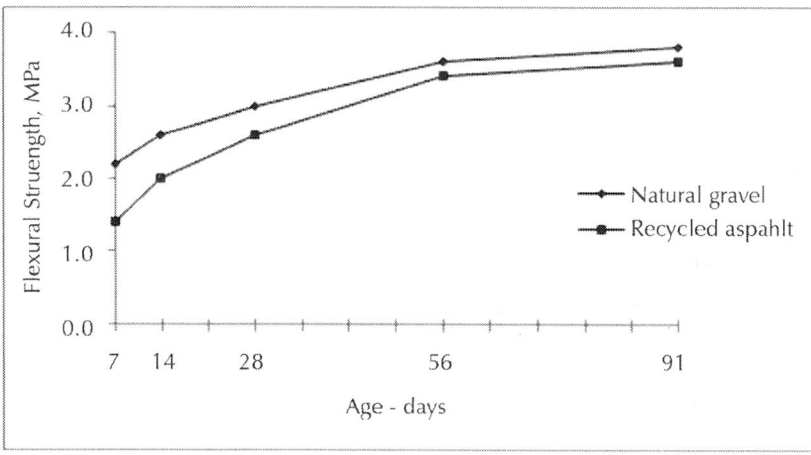

Figure 4.c: Water/cement ratio 0.70.

Figure 4: Relationship between flexural strength and age (mix 1:3:6).

CONCLUSIONS

From the laboratory study carried out to evaluate the performance of recycled asphalt pavement as coarse aggregate in concrete, the following main conclusions can be drawn:

1. RAP aggregate has lower specific gravity and water absorption than the natural aggregate.

2. The crushing and impact value tests as prescribed in BS 812 for assessing the strength of aggregate should not be used in assessing the strength of RAP, a more appropriate assessment method is required.

3. RAP concrete is less workable than the corresponding concrete produced with natural gravel aggregate.

4. The compressive and flexural strength of concrete produced with RAP as coarse aggregate were found to be lower than those made from natural aggregate.

5. The strength of RAP concrete is dependent on the bond strength of the asphalt-mortar coating on the aggregate

6. The maximum compressive strength of concrete that can be produced using RAP as coarse aggregate is approximately 25 MPa.

7. On the basis of this investigation, it is apparent that recycling of waste asphalt pavement for concrete aggregate is feasible and may become a viable and routine process for the generation of aggregate for middle and low strength concrete.

REFERENCES

1. Wainwright P.J., Boni S.K., Some properties of concrete containing sintered domestic refuse as a coarse aggregate, Magazine of Concrete Research, 1983, 35(123), p. 75-85.

2. Okafor F.O., Palm Kernel Shell as a Lightweight Aggregate for Concrete, Cement and Concrete Research, 1988, 18(6), p. 901-910.

3. Mayfield B., Leuli N., Properties of Palletized Blast Furnace Slag Concrete, Concrete Research, 1990, 42(15), p. 29-36.

4. Ravindrarajah R., Loo Y.H., Tam C.T., Recycled Concrete as Fine and Course Aggregate in Concrete, Concrete Research, 1987, 39(141), p. 214-220.

5. Ryu J.S., An Experimental Study on the Effect of Recycled Aggregate on Concrete Properties, Magazine of Concrete Research, 2002, 54(1), p. 7-12.

6. Evangelista L., de Brito J., Mechanical Behaviour of Concrete Made with Fine Recycled Concrete Aggregates, Cement and Concrete Composites, 2007, 29(5), p. 397-401.

7. Tam V.W.Y; Wang K., Tam C.M., Ways to facilitate the use of Recycled Aggregate Concrete, Proceedings of the ICE – Waste and Resource Management, 2007, 160(3), p. 125-129

8. Kou S.C., Poon C.S., Mechanical Properties of 5-years-old Concrete prepared with Recycled Aggregate Obtained from Three Different Sources, Magazine of Concrete Research, 2008, 60(1), p. 57-64.

9. Ann K.Y.; Moon H.Y.; Kim Y.B., Ryou J., Durability of Recycled Aggregate Concrete using Pozzolanic Materials, Waste Management, 2008, 28(6), p. 993-999.

10. Kou S.C., Poon C. S., Properties of Self-compacting Concrete Prepared with Coarse and Fine Recycled Concrete Aggregate, Cement and Concrete Composites, 2009, 31(9), p. 622-627.

11. Li G., Zhao Y., Pang S.S., Huang W., Experimental Study of Cement-asphalt Emulsion Composite, Cement and Concrete Research, 1998, 28(5), p. 635-641.

12. Huang B., Shu X., Li G., Laboratory investigation of Portland Cement Concrete Containing Recycled Asphalt Pavements, Cement and Concrete Research, 2005, 35(10), p. 2008-2013.

13. BS 12: Specification for Portland Cement", British Standard Institution, London, 1996.

14. BS 882: Part 2, Specification for Aggregate from Natural Sources for Concrete, British Standard Institution, London, 1983.

15. BS 812: Parts 1, 2 and 3, Methods of Sampling and Testing Mineral Aggregate, Sand and Fillers, British Standard Institution, London, 1975.

16. BS 1881: Part 102, Methods for Determination of Slump", British Standard Institution, London, 1983.

3

Effects of Diurnal Temperature Dynamics on Curing of Cold-Emulsion Reclaimed Asphalt Pavements

Kiplagat Chelelgo

School of Civil Engineering, University of Nottingham, UK

ABSTRACT

Strength development in Cold-Emulsion Reclaimed Asphalt Pavements is gradual and largely dependent on the rate at which curing proceeds. Its early life strength is therefore low and presents a major challenge in material specification for mechanistic pavement design. The solution has been to subject a sample of the mixture to be used in the pavement to accelerated laboratory curing to the attainment of Equilibrium Moisture Content (EMC) condition. Fatigue and Stiffness parameters of the mix along with the chemical properties of the binder can be determined from the cured samples and results incorporated into the pavement design process. The emphasis is in the choice of a laboratory curing protocol that adequately simulates expected curing trends in the field. Protocols in popular use employ steady state curing temperatures to predict long term behaviour of Reclaimed Asphalt Pavements. This project set out to investigate the likely effects of seasonal variations and diurnal cycles in ambient temperatures on the engineering properties of Reclaimed Asphalt and the incorporated binders. To simulate the above phenomena, a predictive model was adopted in computation of high and low temperature peaks that can be expected in two pavements, one in the tropics and the other in a temperate region. The resulting sets of temperatures were used to cyclically cure Reclaimed Asphalt Pavement cores that were manufactured by artificially aging Dense Bitumen Macadam (DBM) in the laboratory and mixing it with a cationic bitumen emulsion. Another set of cores were subjected to steady temperatures as is the current practice. This acted as a control for the two cyclic temperatures under study. Use was made of a suite of tests available in the Nottingham Asphalt Tester (NAT) to determine stiffness and fatigue properties of the mix under the three treatments. Bitumen binders recovered at the end of curing was tested for penetration, softening point temperatures and percentage of Asphaltenes. The findings pointed at a likelihood of severe treatment of asphalt samples by the existing laboratory curing protocols. Curing at a steady temperature led to a lower fatigue life, over estimation of early life strength and underestimation of long term strength of the cold asphalt. Low penetration values, high softening point temperatures and high percentages of Asphaltenes in bitumen from the cured samples attest to severe aging of the samples.

INTRODUCTION

Pavement recycling as a means of rehabilitating distressed rigid and flexible pavements has continued to grow in popularity due to the environmental and economic benefits it brings. Materials from the two major types of pavements can successfully be recycled, although recycling of flexible pavements dominates. This is due to its popular use in paved areas. To reliably popularize pavement recycling as an alternative to conventional rehabilitation techniques, the performance capabilities of recycled materials should be easy to characterize and specify.

Pavement recycling falls into the categories of Hot Recycling (HR) and Cold Recycling (CR). Hot recycling uses the same technology as Hot Mixed Asphalt and therefore has the disadvantage of huge energy consumption and associated high gaseous emissions. Cold recycling, on the other hand, utilizes softer grades of bitumen, bitumen cut-backs, foamed bitumen or bitumen emulsions. Foamed bitumen and bitumen emulsions have gained favour owing to their good Health Safety and Environment (HSE) record [1].

Whereas it has an excellent environmental record, Cold-Emulsion Reclaimed Asphalt Pavement has a low early-life strength which has resigned its use to low volume roads and works in remote areas where strength is not a key requirement. This, however, has not lessened the popularity of emulsion-reclaimed asphalt pavements. They stand out in addressing reflection cracking, which is a major form of distress suffered by pavements with stiff bases.

Curing is the phenomenon that controls the rate of strength gain in Cold-Emulsion Reclaimed Asphalt Pavements. The role of curing is twofold. One is to rid the mix of moisture so as to allow for direct contact between aggregates and the binder and the second is, to activate the aged binder in Reclaimed Asphalt Pavements via the process of 'fluxing'.

The project set out to study the effects of alternate heating and cooling resulting from seasonal and diurnal pavement temperature cycles on the rate of curing, strength development and fatigue properties of Cold-Emulsion Reclaimed Asphalt mixtures. The aim was to assess the suitability of the existing protocols in mimicking on-site conditions without exaggerating the aging process in bitumen.

REVIEW OF LITERATURE

Trends in Pavement Recycling

Use of pavement recycling as an alternative to the conventional pavement rehabilitation methods started on a small scale in the 1930's and gained much of its popularity in the 1970's due to the energy crisis that hit the globe in 1973 [2]. The resulting fuel associated rise in construction costs spark a research interest into viability of pavement recycling starting in Europe, Australia, United States and South Africa. Laboratory models and field trial sections were constructed and monitored over time to ascertain the engineering properties of the recycled asphalts and to devise ways of improving their performance to match those of conventional rehabilitation materials.

Curing Protocols

Researches by individuals and agencies have adopted a number of accelerated laboratory curing protocols in an attempt to estimate service life of cold mixes. All protocols encountered in the course of this study use steady state temperatures and a few are cited hereunder.

- 3 days cure at 60°C corresponding to the construction period and early field life of the mix, i.e. up to 1 year in the field. [3]
- 14 days cure at 35°C and Relative Humidity of 20% corresponding to between 1 and 3 years in the field in the temperate regions and under low to medium traffic volumes [4, 5, 6]
- 14 days at 18°C at Relative Humidity of 50% to simulate short term curing (a few weeks after laying) in temperate region. [5]
- Curing in the mould for 24 hours at ambient temperatures followed by 48 hours curing at 40°C to simulate 6 months in the field [7]

Two cyclic temperature protocols were developed by adopting pavement temperature models that have hitherto found use in specification of performance grade bitumen and in determination of pavement stiffness parameters for back-calculations in Falling Weigh Deflection (FWD) tests [8]. Asphalt pavements are subjected to cycles of heating and cooling in response to seasonal temperature variations and diurnal temperature cycles. A combination of factors such as solar radiation, air temperature, pavement reflectance, precipitation, freezing-thawing cycles alongside other physical and environmental conditions act to influence the temperature dynamics in the pavement.

Superpave (Superior Performing Asphalt Pavements) under the Strategic Highway Research Program (SHRP) developed a simple algorithm for computation of asphalt pavement temperatures at various depths below the asphalt surface [9]. The model uses ambient temperatures and latitude data to compute pavement temperatures at the surface and depths below the surface, at any point in the globe. For the purpose of this study, two temperature peaks for two cities, one in the tropics and the other in the temperate environment, were computed using the maximum and minimum air temperature using equations (1) and (2) below. The driest and warmest periods were chosen as being the appropriate for laying asphalt concrete in both cities. That falls between June and October in Nairobi and the period between Mid-May and Mid-August in London.

$$T_{Surf} = T_{Air} - 0.00618Lat^2 + 0.2289Lat + 24.4 \tag{1}$$

Where,

T_{Surf} = Temperature at the surface (°C)

T_{Air} = Ambient Temperature (°C)

Lat = Latitude of the region concerned (Degrees)

For temperatures at different depths, the relationship below is used.

$$T_d = T_{Surf}(1 - 0.063d + 0.007d^2 - 0.004d^3) \tag{2}$$

Where,

T_d = Temperature at depth d (°F)

T_{Surf} = Temperature at the surface (°F)

d = Depth from the surface (inches)

Fatani et al [10] conducted a study on pavement temperatures in Saudi Arabia and found out that the maximum temperatures in flexible pavements are recorded at depths of 20mm below the pavement surface. That is approximately halfway through a typical pavement surfacing and can logically represent the average conditions in the pavement.

Climatic data obtained from *BBC Weather* were used to compute the average minimum and maximum air temperature as well as the number of sunshine hours [11, 12]. The resulting values of the upper and lower temperature peaks for the tropical and temperate regions were 44°C and 34°C and 37°C and 29°C respectively. The protocol proposed by Asphalt Institute i.e., 14 days at 35°C [4] was adopted as the steady state curing temperature. Three thermostatically controlled conditioning cabinets were used to apply the chosen curing temperatures in the laboratory.

Table 1: Minimum and maximum pavement temperatures.

Location	Latitude (Degrees) [13]	T Air		T Surface				T 20mm			
		Min	Max	Min: °C	/°F	Max: °C	/°F	Min: °F	/°C	Max: °F	/°C
Nairobi	1.27	11.6	22.2	36.3	97.3	46.9	116.4	92.9	33.8	111.1	43.9
London	51.5	11.8	20	31.6	88.8	39.8	103.6	84.8	29.3	98.9	37.2

MATERIALS, EQUIPMENT AND METHODS

Materials

The research utilized 60 asphalt cores of average dimensions 100mm diameter and 50mm height manufactured in the laboratory using aged Dense Bitumen Macadam of granite origin, dust, mineral filler and a cationic bitumen emulsion. The focus was on cold emulsion reclaimed asphalt pavement fit for use as surfacing and thus aggregates of maximum size 20mm were used.

Aggregates

The aggregates used in this research were derived from artificially aged Dense Bitumen Macadam (DBM) obtained from Cliffe Hill Quarry in Leicester. Dust and the Filler were also obtained from the same source. The residual binder in the DBM after being kept at ambient temperature for close to a month was determined as 4.25% by mass of aggregates. Its penetration ranged between 20dmm and 21dmm. To simulate aging, the DBM was reheated to 160°C, laid in slabs of dimensions 305mm by 305mm by 50mm and allowed to cool for two days before being crushed and tested for residual bitumen. The penetration of bitumen had dropped to 14 dmm signifying substantial aging.

Bituminous Binder

A cationic emulsion containing 60% bitumen and 40% water was used as the binder in the preparation of the cold mix. The bitumen emulsion of Venezuela origin was supplied by Nynas Asphalts, UK. At the time of supply, the supplier reported the penetration of the emulsion as being 48dmm and its softening point as 51.4°C. The bitumen emulsion constituted 6% by mass of dry aggregates. Pre-wetting water constituting 1.5% by mass of dry aggregates was incorporated to disperse the emulsion besides improving the workability.

Equipment

To monitor performance parameters of the three sets of specimens as curing proceeded, use was made of a suite of tests available in the Nottingham Asphalt Tester (NAT) [14]. NAT was the main piece of equipment but other equipment that came in handy in preparation and conditioning of samples included a Jaw Crusher, Sieve Shaker, Hobart Mixer, Shear Gyratory Compactors and Conditioning Cabinets.

Shear Gyratory Compactor was used to manufacture test cores in the laboratory by simulating the kneading action of rollers used to compact asphalt on site.

Mix Design and Specimen Preparation

Reclaimed Asphalt Pavement, Dust and Filler were graded separately and blended in proportions of 65%, 30% and 5% respectively to produce an overall gradation falling within the envelope defined by the lower and upper bounds of the Overseas Road Notes No. 19 and 31 [15, 16]. Several proportions were tried with the aim of approaching the maximum dry density curve as defined by Cooper Equation below [17].

$$P= \frac{(100-F) (d^n-0.075^n)}{D^n-0.075^n} + F \quad (3)$$

Where,

P = Percent material passing sieve size d (mm)

D = Maximum aggregate size (mm)

F = Percent filler (%)

n = Exponent that defines the curvature of the gradation curve, usually 0.45 for maximum packing of particles [18].

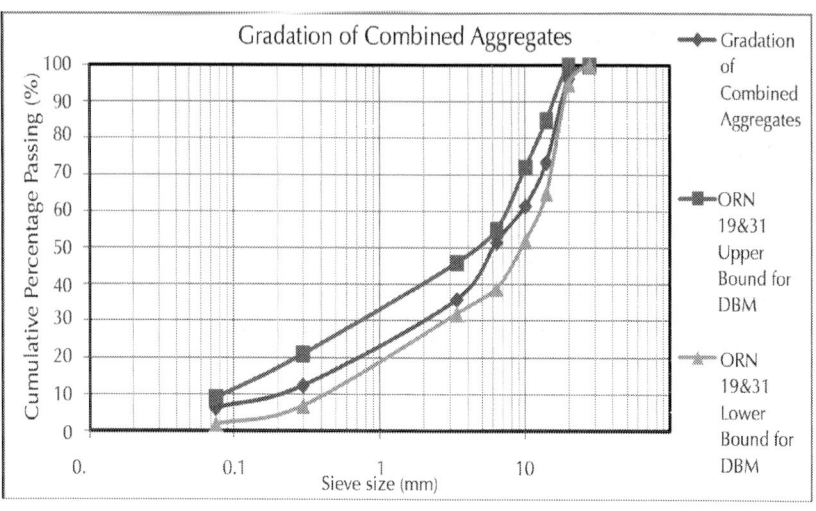

Figure 1: Aggregate gradation.

Cylindrical specimens for laboratory curing and testing were prepared based on the determined mixed aggregates gradation, emulsion requirement and the optimum pre-wetting water content. The target dimensions for both the ITFT and ITSM tests were 100mm diameter and 50mm height. The aggregates were batched into metallic cans and conditioned overnight in a conditioning cabinet set at 35°C to drive out any moisture.

The aggregates were then mixed in a Hobart mixer with addition of 1.5% water and 6% bitumen emulsion by mass of dry aggregates. The resultant mix was then weighed into steel moulds and compacted in the gyratory compactor to the target density. The cylinders were cured at room temperature for 24 hours before being transferred into the respective thermostatically controlled conditioning cabinets where weight loss was tracked as a means of evaluating curing progression.

Testing Methods

Indirect Tensile Stiffness Modulus Test (ITSM)

Stiffness of an asphalt mix is a reflection of its ability to effectively spread tyre loads to the underlying pavement layers without damaging the foundations.

Figure 2: NAT in the ITSM configuration.

Stiffness tests in the laboratory can be performed either by the Uniaxial Test, Indirect Tensile Test or the Beam Tests. In this exercise, stiffness test was done in accordance with BS DD 213: 1993, which specifies the method of performing Indirect Tensile Stiffness Modulus Test in the Nottingham Asphalt Tester

Indirect tensile fatigue test (ITFT)

Fatigue is the structural damage suffered by a material when subjected to a cyclic or repeated stress that is generally of magnitude below the ultimate tensile strength of the material. Traffic and thermal loads in a pavement induce alternate stretching and relaxation in the binder matrix which eventually leads to fracture being manifested as fatigue cracks on the road surface.

Determination of fatigue life in the laboratory can be done using simple flexure, uniaxial test or the indirect tensile test. The latter is preferred due to ease in specimen fabrication and is the method adopted in the draft specification - BS DD ABF: 2003 and used in this exercise.

Figure 3: NAT in the ITFT configuration.

RESULTS AND DISCUSSIONS

Moisture Loss

Moisture loss with time was reckoned from the residual moisture as a percentage of the total mix by mass. Curiously, the three curing regimes displayed closely similar trends in water loss. As can be seen from Figure-4 below, the three protocols followed more less the same trend in moisture loss. 10% of the total water content was lost in the 24 hour period of curing in the mould and another 80% was lost after curing for one day in the condition cabinets. 90% of the total moisture content had been lost on the second day, suggesting that temperature may not be the key player in the evaporation mechanism. Equilibrium Moisture Content (EMC) seemed to have been achieved after 6 days of curing.

Indirect Tensile Stiffness Modulus Test (ITSM)

Stiffness modulus of cylinders from the three curing protocols was determined at six time intervals. The cyclic temperatures started on the lower peaks and the effect was reflected in the stiffness modulus determined after a day of curing in the ovens. That would have been equivalent to on-site laying of cold-emulsion in the evening when the temperatures are low.

Strength development in the cylinders cured at 35°C rose gradually in an almost linear manner while those under cyclic temperatures rose gradually with decreasing gradients towards a peak.

Wide variations in material properties are common in cold mixes but a general trend can be drawn from the results plotted in Figure-5 above. It can be deduced that laboratory curing of cold emulsion asphalts at 35°C closely predicts the intermediate strength, overestimates the early strength and underestimates the long term strength. The tropical conditions show a faster rate of strength development which may generally be underestimated if use is made of the existing protocol of curing at 35°C. A stiffness modulus of 2000 Mpa is sufficient to support low to medium traffic [17], which, in this case, can be achieved from as early as 12 days in the tropical conditions.

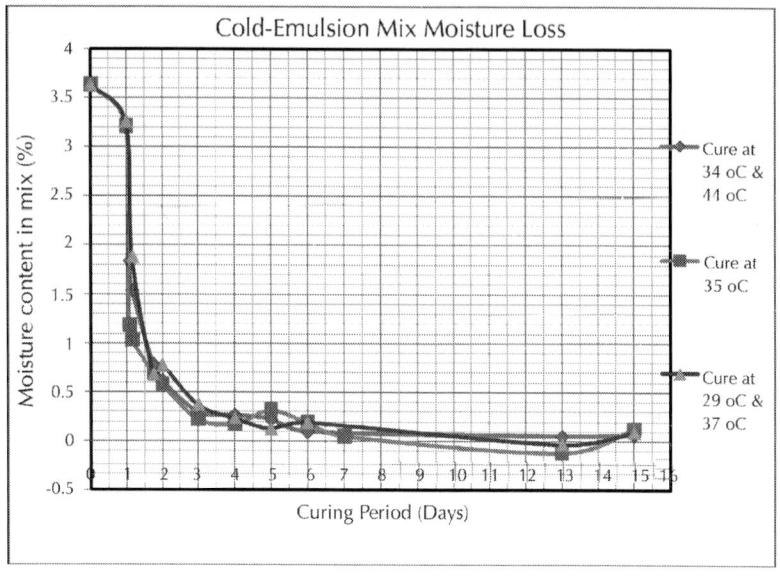

Figure 4: Cold-Emulsion RAP moisture loss.

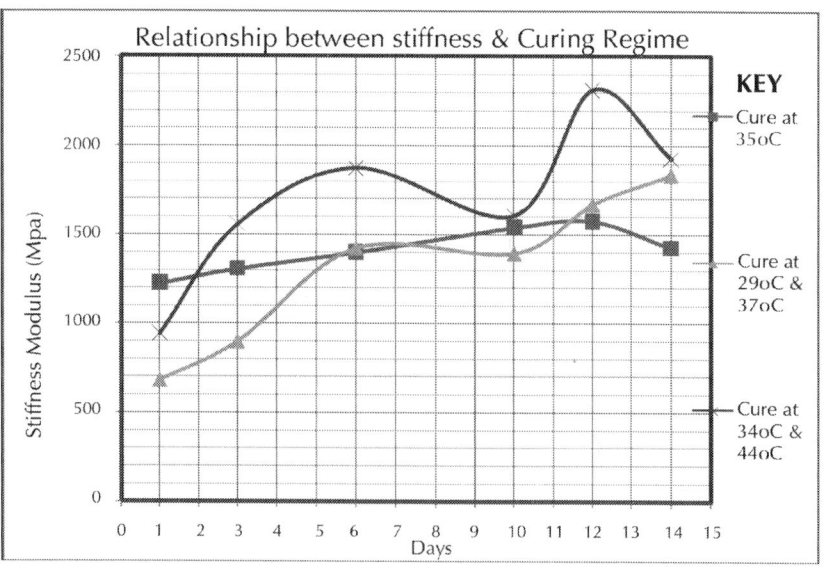

Figure 5: Strength evolution in cold-emulsion RAP.

Indirect Tensile Fatigue Test (ITFT)

The aim of the test is to load the specimen to failure by applying alternate stress or strain, and to determine the number of load applications to cause the failure. Cylinders cured for 10 and 12 days were curing for a further four and two days respectively. This was aimed at providing additional number samples for determination of fatigue life which was targeted at samples cured for 14 days. Ten samples conditioned overnight were subjected to load pulses ranging from 600 kPa to 100 kPa.

Fatigue characteristics are determined by plotting maximum tensile horizontal strain versus life to failure on logarithmic scales. Fatigue characteristics of materials cured under the three regimes were compared by plotting the maximum horizontal tensile strains against the number of cycles to failure and generating fatigue relationships by use of power trend lines. Fatigue life of each mixture was obtained from its linear regression model by assuming logarithmic linearity of fatigue life. Samples cured cyclically at 29°C and 37°C exhibited the highest level of fatigue resistance while those cured at 34°C and 44°C had the lowest life to failure.

Early failure by fatigue cracking in the mix cured cyclically at 34°C and 44°C can be explained by considering the likely effects of high temperatures on the bitumen binder. At temperatures of 44°C, bitumen could be losing the volatile components and thus ending up being brittle.

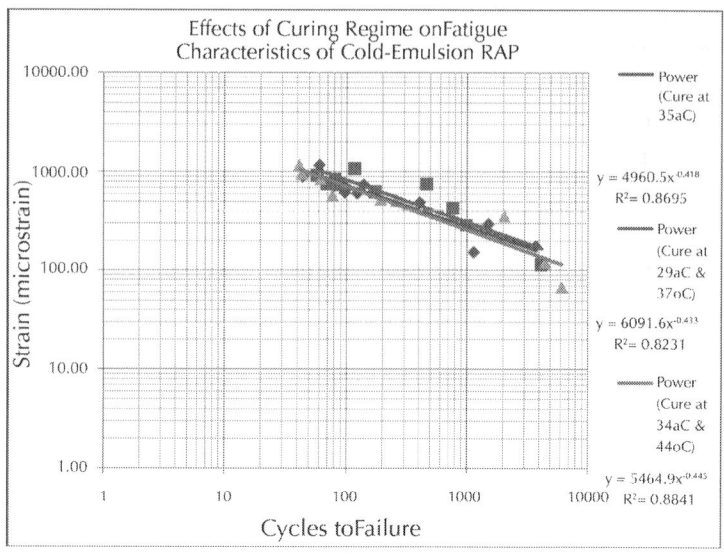

Figure 6: Comparison of fatigue life.

Table 2: Linear regression values

Curing protocol	Equation based on N	N at 100µε $_f$	R^2
35°C	$y = 4960.5x^{-.418}$	11,384	0.87
29°C and 37°C	$y = 6091.6x^{-.433}$	13,237	0.82
34°C and 44°C	$y = 5464.9x^{-.445}$	8,026	0.88

Properties of Recovered Binders

Bitumen was recovered from the cores before and after being subjected to the three curing conditions. Table-3 below presents the characterization of the binders in terms of Penetration, Softening Point and percentage of Asphaltenes. These three parameters were used to assess severity of aging of bitumen caused by the curing regimes. The results indicate that curing at the steady temperature of 35°C results in the highest degree of aging as compared to the other two regimes. The increase in Asphaltene contents after curing is partly due to activation of bitumen in the Reclaimed Asphalt Pavement.

It is therefore logical to suggest that curing at the steady temperature of 35°C to simulate on site conditions may be too harsh a treatment for cold-emulsion Reclaimed Asphalt Pavement used in both the temperate and tropical conditions.

Table 3: Properties of recovered binders

		Penetration (dmm)	Softeningpoint (°C)	Asphaltenecontent (%)
Before curing		26	61	14.8
After curing	35°C	21	64	16.3
	29°C and 37°C	32	59.2	16.2
	34°C and 44°C	22	63	16.1

CONCLUSION AND RECOMMENDATIONS

The study of the fundamental properties of the materials cured under three curing protocols, two cyclic and one steady state, revealed a potentially useful correlation between laboratory curing temperatures and on-site curing. The results from the study form a basis for a detailed study into the precise behaviour of cold asphalts under different site conditions. Based on the results of the laboratory tests, the study makes the following tentative conclusions.

- Prevailing pavement temperature has a greater effect on the rate of 'fluxing' or activation of aged binder in a Cold-Emulsion Reclaimed Asphalt Pavement than it has on the rate of moisture loss.
- Steady temperature laboratory curing regimes severely age the binder in the Cold-Emulsion Reclaimed Asphalt Pavement mixtures.

Recommendations

The study gives an indication of the behaviour of Cold-Emulsion RAP though extensive monitoring of actual field performance still needs to be done to validate the results. A study incorporating more variables that interactively influence the curing process needs to be conducted. The study makes the following recommendations for future research:

- Laboratory generated cores should be introduced in one go into the conditioning cabinets to prevent absorption of moisture by the already cured samples.
- Effects of humidity and air draught be incorporated in laboratory curing to give a better simulation of the on-site conditions.
- Cyclic temperatures in the laboratory should be applied stepwise to better simulate the diurnal temperature cycles.

ACKNOWLEDGEMENTS

I would like to extend my sincere gratitude to the University of Nottingham for the financial support offered through the "Developing Solutions for Africa Scholarship Scheme". It is through the fund that I was able to enroll for the M.Sc. program under which this research was undertaken.

Many thanks also go to my supervisor, Prof. Tony Parry of the University of Nottingham for his visionary guidance and technical advice throughout the duration of the project. His able leadership kept the project on course.

Last but not least are the members of my research team, Oluwaseyi, Cheng, Chibuzor and Dunshun. They greatly challenged my thinking in many aspects of the project and they deserve recognition for the all the knowledge that they freely shared.

REFERENCES

1.   Transport Research Board. 2006. Asphalt Emulsion Technology. Circular No. E-10.

2.   Van Wijk A.J. *et al.* 1984. Construction of Cold Recycled Pavements Using Emulsion as a Binder. Proceedings of the Fourth Conference on Asphalt Pavements for Southern Africa. Cape Town, South Africa.

3.   Ruckel P.J. *et al.* 1982. Foamed-Asphalt Paving Mixtures: Preparation of Design Mixes and Treatment of Test Specimens. Asphalt Materials. TRL.

4.   British Standards Institute. 2006. BS 4324-2:2006, Bitumen Road Emulsions-Part 2: Code of Practice for the Use of Cationic Bituminous Emulsions on Roads and other Paved Areas.

5.   Serfass J.P. *et al.* 2004. Influence of Curing on Cold Mix Mechanical Performance. Materials and Structures. 37: 365-368.

6.   Brennan M.J *et al.* 2007. Laboratory Performance of Emulsion-Bound Macadam Manufactured Using Different Production Processes. Asphalt Professional. Issue No. 29.

7.   Kekwick S.V. 2005. Best Practice: Bitumen-Emulsion and Foamed Bitumen Materials in Laboratory Processing. South African Transport Convention (SATC).

8.   Yavuzturk C., Ksaibati and Chiasson A.D. 2005. Assessment of Temperature Fluctuations in Asphalt Pavements Due to Thermal Environmental Conditions Using a Two-Dimensional, Transient Finite-Difference Approach. Journal of Materials in Civil Engineering. 17: 465-475.

9.   Mallick B and El-Korchi T. 2009. *Pavement Engineering: Principles and Practice.* Taylor and Francis Group, USA.

10. Fatani M. *et al.* 1994. Evaluation of Permanent Deformation of Asphalt Concrete Pavements in Saudi Arabia. Final Report Submitted to King Abulaziz City for Science and Technology (KACST).

11. BBC Weather Centre. 2009. *Average Conditions for* Nairobi, Kenya Available at: http://www.bbc.co.uk/weather/world/city_guides/results.shtml?tt=TT000300 [Accessed on 10 April 2009].

12. BBC Weather Centre. 2009. Average Conditions for London, United Kingdom. Available at: http://www.bbc.co.uk/weather/world/city_guides/results.shtml?tt=TT003790 [Accessed on 10 April 2009]. Assignment: The World. 2003. Finding Places

with Longitude and Latitude. Available at: http://atwonline.org/latitude.swf [Accessed on 10 April 2009].

13. Sri Sunarjono M.T. 2008. The Influence of Foamed Bitumen Characteristics on Cold-Mix Asphalt Properties. PhD Thesis, University of Nottingham.

14. Department for International Development-DFID. 2002. A Guide to the Design of Hot Mixed Asphalt in Tropical and Sub-Tropical Countries. Overseas Road Note No.19, TRL, UK.

15. Overseas Development Agency. 1993. A Guide to the Structural Design of Bitumen-Surfaced Roads in Tropical and Sub-Tropical Countries. Overseas Road Note 31, 4th Ed. TRL, UK.

16. Thanaya N.A. 2007. Evaluating and Improving the Performance of Cold Asphalt Emulsion Mixes. Civil Engineering Dimension. 9(2): 64-69.

17. US Department of Transportation, Federal Highway Administration. 2001. FHWA Publication No. FHWA-RD-01-052: Superpave Mixture Design Guide. Washington DC.

Chapter 4

Comparative Structural Analysis of Flexible Pavements Using Finite Element Method

[1]Ankit Gupta, [2]Abhinav Kumar

[1]Assistant Professor, Department of Civil Engineering, NIT Hamirpur, India.

[2]Former Graduate Student, Department of Civil Engineering, NIT Hamirpur, India

ABSTRACT

The evaluation of bituminous concrete mixes for their tendency to rutting has been an important research field for many years. Rutting is a major type of distress encountered in bituminous pavements. The Finite Element Method (FEM) is a numerical analysis technique to obtain various structural parameters such as stress, strain and deflection of pavement layers. The objective of this paper is to study the sensitivity of these variables in reducing the vertical surface deflections, the critical tensile strains at the bottom of the bitumen layer and the critical compressive strains on the top of subgrade using the finite element method. This study has been carried out in order to compare the performance of flexible pavement using the finite element method and KENLAYER. Vertical surface deflections in flexible pavements have always been a major concern and are used as a criterion for pavement design. It is desirable to reduce the deflections as much as possible. This paper deals with ways to reduce deflections by varying the design configuration, such as increasing the Hot Mix Asphalt (HMA) modulus, the base modulus, sub base modulus and the subgrade modulus. Another objective of the present study is to investigate the effectiveness of two different methods in reducing vertical surface deflections (w_o) and the critical tensile strains in the bitumen layer (ε_t) or the radial strains at the bottom layer of HMA. The finite element method was adopted to evaluate the effectiveness of the two methods and the sensitivity of various factors.

INTRODUCTION

Structural analysis in pavements has been greatly developed since the initial studies carried out by Boussinesq in which soils were modeled as a linear-elastic material (Boussinesq, 1885). Boussinesq's theory was then extended to multilayer elastic models due to the work of Burmister (Burmister, 1945) and Schiffman (Schiffman, 1962). Rutting is caused by the accumulation of permanent deformation in all pavement layers under repetitive traffic loading. Among the contributors of rut depth in the different pavement layers, the cumulative permanent deformation in the surface course of bituminous pavement is known to be responsible for a major portion of the final rut depth measured on the pavement surface. Thus, rutting occurs only on flexible pavements, as indicated by the permanent deformation or rut depth along the wheel paths. The width and depth of the rut are widely affected by structural characteristics of the pavement layers (thickness and material quality), traffic loads and environmental conditions (Huang, 1993). The numerical analysis of the pavement layer is based on the finite element method (FEM). Figure 1 represents a cross section of a basic modern pavement system, showing its major components.

Figure 1: Basic components of a typical pavement system.

This paper deals with different possible ways to reduce vertical surface deflection by varying the design configuration (input parameters), such as increasing hot mix asphalt (bitumen) modulus, the base modulus, the sub-base modulus and the subgrade modulus. The primary objective of this study is to analyze the sensitivity of the layer modulus variables in reducing the surface deflection and the soil stress in flexible pavement.

Rutting due to permanent deformation is considered one of the most serious distress mechanisms in bituminous pavements. It leads to traffic hazards by affecting vehicle steering. Furthermore, an impervious road surface will trap water, snow and ice that cause hydroplaning and loss of friction. Longitudinal cracks sometimes occur in deep ruts where they drain free water into the underlying pavement layers, thereby increasing the deterioration rate. The factors affecting permanent deformations can be divided into traffic loading, material properties and climatic conditions. Modeling is a valuable tool used for pavement design and residue assessment. The first pavement deterioration models were entirely empirical but mechanistic principles have been introduced in recent years (Gupta et al., 2014).

We have employed the ANSYS version 11 and the KENLAYER programs/software packages for the purpose of modeling and analysis of the flexible pavement subjected to repetitive wheel load. ANSYS is a finite element numerical technique and a mechanistic approach analysis, while KENLAYER is an empirical analysis technique for pavements. The finite element method, its practical application often known as finite element analysis (FEA) is a numerical technique for finding approximate solutions to Partial Differential Equations (PDE) and their systems. FEM is a special case of the more general Galerkin method with polynomial approximation functions. The solution approach is based on eliminating the spatial derivatives from the PDE. This approximates the PDE with a system of algebraic equations for steady state problems and a system of ordinary differential equations for transient problems. These equation systems are linear if the underlying PDE is linear, and vice versa. Algebraic equation systems are solved using numerical linear algebra methods. Ordinary differential equations that arise in transient problems are then numerically integrated using standard techniques such as the Euler's method or the Runga-Kutta method. Premature failure in flexible pavement has long been a problem in many roads with the large increase in truck axle load. To fully utilize each pavement material in a costefficient manner, a pavement should generally have a design, striking a reasonable balance between the rutting and fatigue modes of distress. The purpose of this paper is to develop an approach for achieving an economic, balanced and quality based evaluation of the various components of the flexible pavement. The methodology is based on the damage analysis concept which has been performed to evaluate rutting on different pavement moduli and Poisson's ratio by using the ANSYS and KENLAYER programs.

There are various modes of failure of flexible pavement. Flexible pavement is constructed always bearing in mind its durability, and surface skid resistance under in service conditions. Further it is expected to exhibit minimum possible cracking and rutting in flexible pavement layers. Large stresses and strains are produced with thicker layers carrying higher flexural stress than thinner layers, while subjected to large and more concentrated loads. The increased rutting or decreased fatigue life of the flexible pavement may be attributed to the shortcomings of the application of the flexible pavement analysis and the lack of attention to identify the pavement components which aid in achieving a balanced section which renders equal pavement lives with respect to rutting and fatigue. The use of FEM model through ANSYS allows the model to accommodate the load dependent stiffness of the granular and subgrade materials, although most of the models still use linear elastic theory as constitutive relationship.

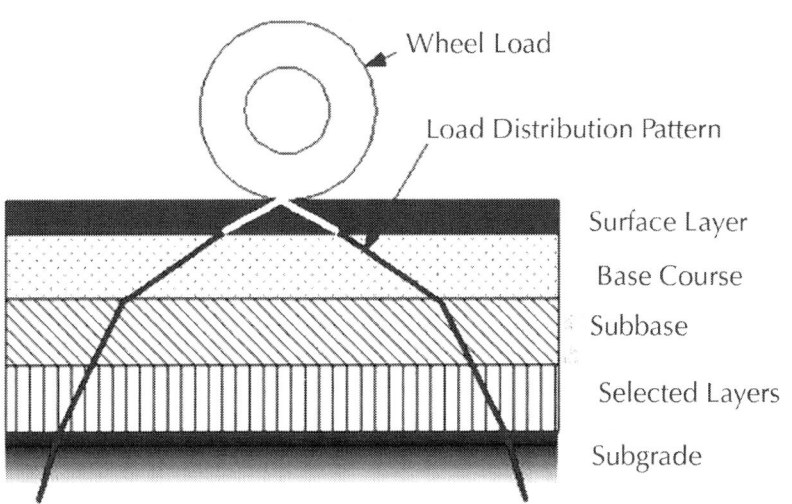

Figure 2: Load distribution along various layers.

In the analysis of flexible pavement, axle loads on the surface of the pavement produce two different types of strains which are believed to be most critical for design purposes. These are the horizontal tensile strains; ε_t at the bottom of the bitumen layer, and the vertical compressive strain; ε_c at the top of the subgrade layer. If the horizontal tensile strain ε_t is excessive, cracking of the surface layer will occur and the pavement will fail due to fatigue. If the vertical compressive strain ε_c is excessive, permanent deformations are observed at the surface of the pavement structure (from overloading the subgrade) and pavement fails due to rutting.

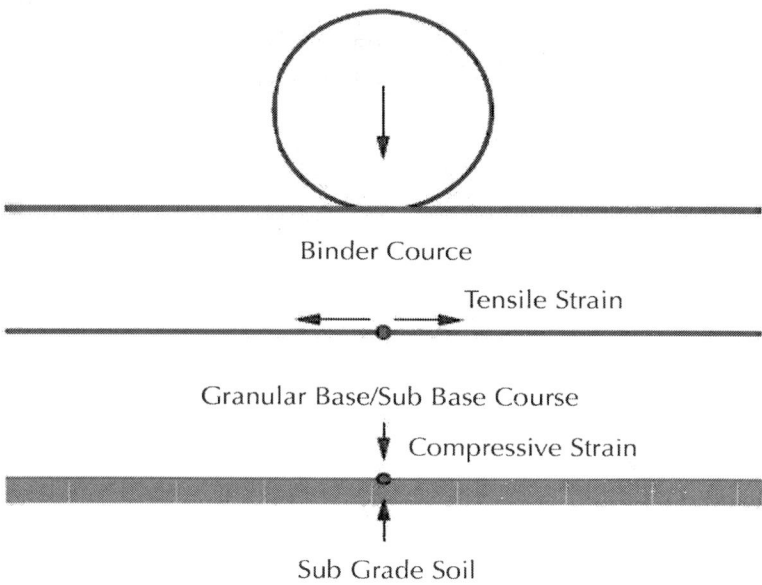

Binder Cource

Tensile Strain

Granular Base/Sub Base Course

Compressive Strain

Sub Grade Soil

Figure 3: Failure modes and critical strains

Rutting Failure Criteria

The relationship between rutting failure and compressive strain at the top of subgrade material has been investigated and definitions suggested by various institutions, organizations and individual researchers depending upon the varying load application and pavement material characteristics. The present study employs a model suggested by the Asphalt Institute (Asphalt Institute, 1982), which can be stated as follows:

$$Nr = 1.365 \times 10^{-9}(1/\varepsilon_c)4.477 \quad (1)$$

Where, N_r = number of load repetitions to limit rutting

ε_c = vertical compressive strain at the top of the subgrade

Literature Review

Abed and Al-Azzawi (2012) concluded that the stress level decreased by 14% in the leveling course and 27% in the base course, while the rut depth increased by 12% and 28 % in those respective layers because the material properties had been changed. The modulus of elasticity for surface layer was taken as 2689 N/mm² whereas for base course was taken as 1655 N/mm². Oscarsson and Popescu (2011) concluded that the results from the semi-rigid pavement section indicated that shear stress and elastic shear strain may be difficult to relate to flow rutting in very stiff pavement sections.

METHODOLGY

A typical cross section consists of a bituminous layer with thickness d1 = 100 mm and elastic modulus of E_1 = 229.8 MPa, a base layer with thickness d_2 = 300 mm and elastic modulus E_2 = 114.9 MPa, and a sub base layer with thickness d_3 = 300 mm and elastic modulus E_3 =46 MPa, resting on a subgrade with modulus of elasticity E_4 = 5.74 MPa. This is regarded as a section with reference components. Different likely cross sections that may be used in Indian Roads are considered for analysis through varying the reference components. In other words, E_1 is varied from 229.8 to 1149 MPa, while E_2 from 114.9 to 1200 MPa and E_3 from 46 to 1100 MPa and E_4 from 5.74 to 200 MPa. Materials in each layer are characterized by a modulus of elasticity (E) and a Poisson's ratio (v). Poisson's ratio (v), the values of 0.35, 0.30, 0.30 and 0.40 are considered for bituminous layer, base course, sub base layer and subgrade, respectively. Traffic is expressed in terms of repetitions of single axle load 18-kip applied to the pavement on two sets of dual tires. The studied contact pressure is 0.70 MPa. The dual tire is approximated by two circular plates with a radius of 100 mm and spaced at 350 mm center to center. The detrimental effects of axle load and tire pressure on various pavement sections are examined by computing the tensile strain (ε_t) at the bottom of the bituminous layer and the compressive strain (ε_c) at the top of the subgrade. Subsequently a damage analysis is carried out using the two critical strains to compute pavement life for permanent deformation (rutting). A sensitivity analysis demonstrates the effect of various parameters on flexible pavement. The analysis is performed using the finite element computer package ANSYS. The results indicate that displacements under loading are the closest to mechanistic methods. A research study is then undertaken to incorporate the realistic material properties of the pavement layers and the moving traffic load, in the analysis of the flexible pavement, employing the FEM. For comparison purposes, a flexible pavement is conventionally taken as a multilayered elastic system in the analysis of pavement response, using KENLAYER.

As with models for the prediction of resilient response, there are a large number of models that have been proposed to represent the Permanent Deformation (PD) of granular materials. These models appear to be either based on observed performance or are expressed as a function of the number of load applications/cycles and the applied stress state. Duncan and Chang (1970) proposed a hyperbolic model for predicting plastic strains from triaxial tests as a function of confining and deviator stresses, cohesion, the angle of internal friction and a ratio of compressive strength to an asymptotic stress difference. Well-known geotechnical models of this type are the Cam-Clay (Schofield and Wroth, 1968) and Drucker-Prager (Drucker and Prager, 1952) models. The Mechano-Lattice (ML) method of analysis (Yandell, 1971), determines the elastic and plastic response of the system, as a wheel rolls across the surface of the model. The pavement structure is modeled as a series of springs in a lattice framework. The observation that, after repeated load triaxial testing that materials have a higher secant modulus on unloading than loading is used to develop the plastic strains within the structure. The plastic strains predicted by the ML method are comparable to the measured plastic strains/ruts. Table 1 shows the typical pavement material properties. The material properties are shown on Table 2. A total of 17 cases were analyzed. The finite element mesh is shown on Figure 4. This analysis is based on the assumption that all layers are linearly elastic, although HMA layers are viscoelastic and base layers are nonlinear elastics. Analysis of the pavement model in ANSYS has been carried out with the help of the Drucker-Prager method and regular hexagonal meshing has been used in order to analyze the pavement model at every tiny, infinitesimal element.

As far as the boundary conditions of the pavement are concerned, the subgrade layer has its displacements completely restrained. The sides of the pavement model have no restraints in vertical direction but they are completely restrained over the other two possible displacements, as illustrated on Figure 5. The procedure is performed once the tires have been placed on the bituminous surface. In the contact discretization, the bituminous surface is defined as the master surface, whereas the tire surfaces in contact with the bituminous surface are defined as slave surfaces. Subsequently the simulation is performed and in the equilibrium configuration the results lead to the correct phenomenon.

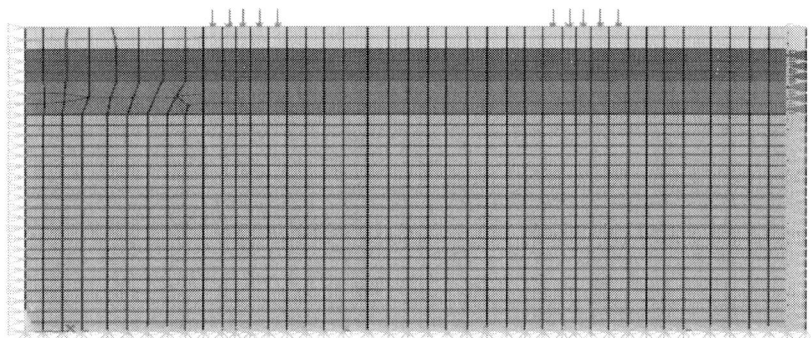

Figure 4: Loading arrangement and meshing

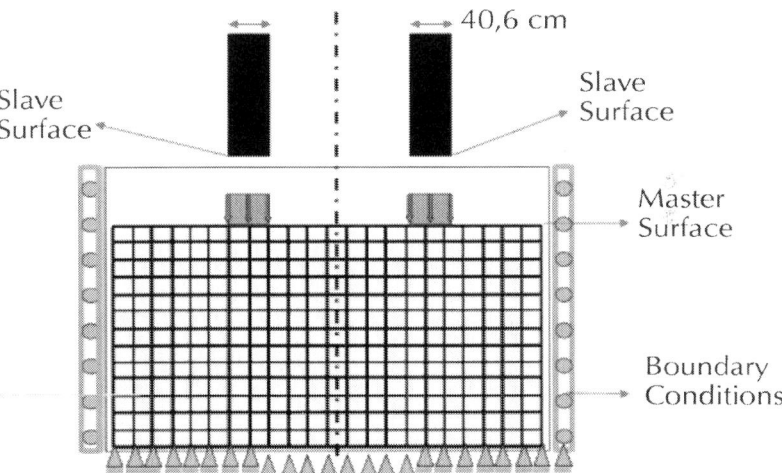

Figure 5: Boundary condition.

Table 1: Pavement material properties

Material	E (MPa)	Poisson's ratio	Unit weight (Kg/m³)
Bituminous surface	229.8	0.35	2400
Soil aggregate base layer	114.9	0.30	2300
Sub base layer	46	0.30	2250
Subgrade layer	5.74	0.40	1800

Table 2: Pavement material modulus range used in analysis

Material	E (MPa)	Poisson's ratio
Bituminous surface	229.8 to 1149	0.35
Soil aggregate base layer	114.9 to 1200	0.30
Sub base layer	46 to 1100	0.30
Subgrade layer	5.74 to 200	0.40

EFFECT OF LAYER MODULUS

This study has been carried out in order to compare flexible pavement performance using FEM and KENLAYER computer programs, respectively. Comparison of the output has been made to determine the governing distress and deterioration models. Table 3 shows the variation of input parameters in the analysis and Figures 6-9 shows the comparative contour plot for the vertical strain of FEM and KENLAYER, respectively. As observed on Figures 6-9, the vertical deflection reduces as the modulus increases at all values of E. It is also noteworthy that, w_0 exhibits no sensitivity with respect to the variation of E_1, as opposed to E_2, E_3 and E_4. The investigated pavement components are elasticity moduli (E_1, E_2, E_3 and E_4) for the bituminous layer, base layer, sub base layer and the subgrade elasticity modulus, respectively. The results of pavement analysis showed that E_4 are the key elements which control the equilibrium between fatigue and rutting lives (N_f and N_r, respectively). This is the case because increasing E_4 sharply increases N_r, and does not affect N_f. The study also concluded that by increasing E_3, E_2 and E_1, N_f and N_r mildly increase. Therefore, it may be stated that E_4 is the most effective component in pavement structure for increasing pavement life, followed by E_3 (high-quality sub base).

Table 3: Variation of input parameters in analysis

CASE	E_1 (Pa)	E_2 (Pa)	E_3 (Pa)	E_4 (Pa)
1	229800,000	114900,000	45960,000	5745,000
2	459600,000	114900,000	45960,000	5745,000
3	689400,000	114900,000	45960,000	5745,000
4	919200,000	114900,000	45960,000	5745,000
5	1149000,000	114900,000	45960,000	5745,000
6	229800,000	229800,000	5745,000	5745,000
7	229800,000	344700,000	45960,000	5745,000
8	229800,000	459600,000	45960,000	5745,000
9	229800,000	850000,000	45960,000	5745,000
10	229800,000	1200000,000	45960,000	5745,000
11	229800,000	114900,000	91920,000	5745,000
12	229800,000	114900,000	183840 000	5745,000
13	229800,000	114900,000	1100000,000	5745,000
14	229800,000	114900,000	45960,000	17235,000
15	229800,000	114900,000	45960,000	34470 000
16	229800,000	114900,000	45960,000	68940,000
17	229800,000	114900,000	45960,000	200000,000

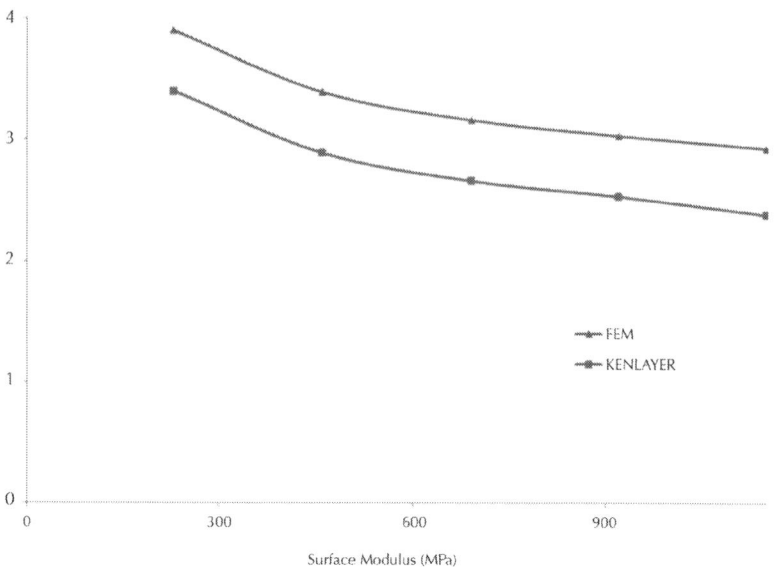

Figure 6: Effect of surface modulus on vertical deflection.

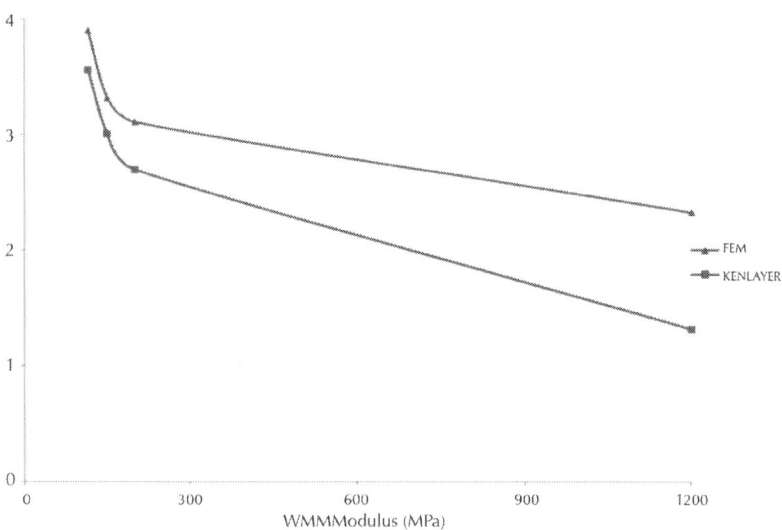

Figure 7: Effect of WMM modulus on vertical deflection.

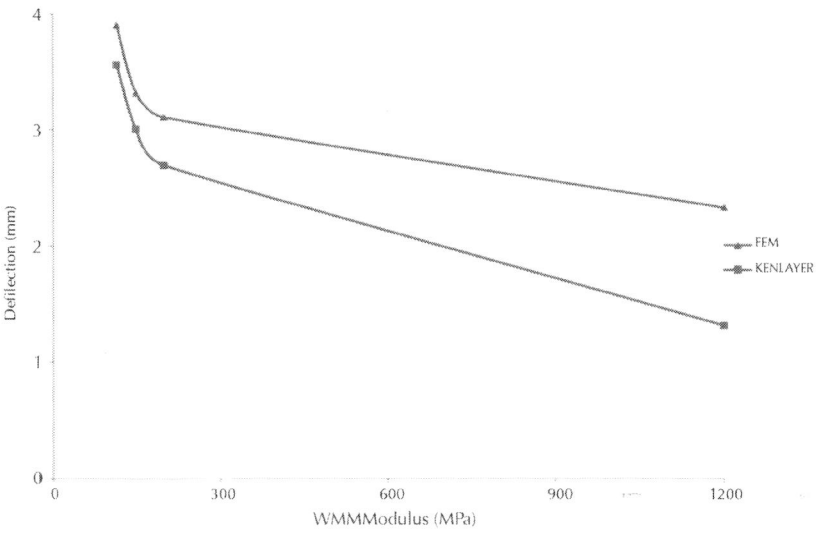

Figure 8: Effect of GSB modulus on vertical deflection.

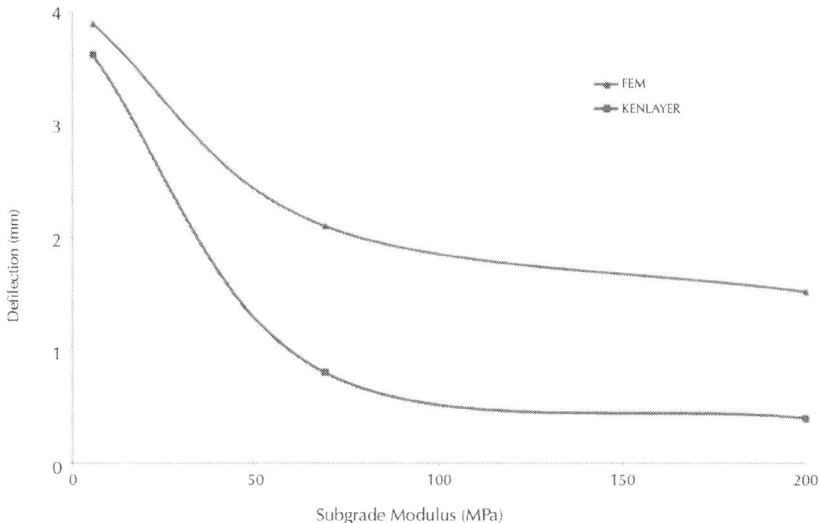

Figure 9: Effect of subgrade modulus on vertical deflection.

CONCLUSIONS

- KENLAYER can be used to predict the performance of flexible pavement more easily and efficiently since it is more user-friendly.
- Subgrade modulus is the key element which controls the excess vertical surface deflection in flexible pavement. Hence, more efforts are required for achieving high value of subgrade modulus as compared to other top layers of pavement.
- Base course and surface layer modulus have minor effects on the excess vertical surface deflection in flexible pavement.
- The design obtained from FEM and KENLAYER analysis is different for the corresponding modulus. However, there are discrepancies in the results obtained from the KENLAYER program. Although the pavement structure was assumed to be linear elastic, calculated maximum vertical deflections are lower than the corresponding results obtained from FEM analysis.

REFERENCES

1. Abed, A.H., and Al-Azzawi, A.A. (2012), "Evaluation of Rutting Depth in Flexible

2. Pavements by Using Finite Element Analysis and Local Empirical Model", American Journal of Engineering and Applied Sciences, Vol. 5, No. 2, pp. 163-169.

3. Asphalt Institute (1982). Research and Development of Asphalt Institute's Thickness

4. Design Manual. 9th Ed., Research Report 82-2, The Asphalt Institute.

5. Boussinesq, J. (1885), "Application des potentiels á l'étude de l'équilibre et du

6. movement des solidesélastique", Paris: Gauthier-Villard.

7. Burmister, D.M. (1945), "The General Theory of Stresses and Displacements in Layered Soil Systems", Journal of Applied Physics. Vol. 16, pp. 89-94, 126-127, 296-302.

8.   Duncan, J.M., and Chang, C.Y., (1970), "Nonlinear Analysis of Stress and Strain inn Soils", ASCE Journal of Soil Mechanics and Foundations Division, 96 (SM5), pp. 1629-1653.

9.   Drucker D.C., and Prager, W., (1952), "Soil Mechanics and Plastic Analysis of Limit Design", Quarterly, Applied Mathematics, Vol.10, No.2, pp. 157-164.

10. Gupta, A., Kumar, P., and Rastogi, R. (2014), "Critical Review of Flexible Pavement

11. Performance Models", Korean Society of Civil Engineers (KSCE), Journal of Civil Engineering, Springer, Vol. 18, No. 1, pp. 142-148.

12. Huang Y.H., (1993), "Pavement Analysis and Design", Englewood Cliffs, New Jersey, Prentice-Hall.Oscarsson, E., and Popescu, L., (2011), "Evaluation of the CalME Permanent Deformation Model for Asphalt Concrete Layers", International Journal of Pavement Research Technology, Vol. 4, No. 1, pp. 21-33.

13.   Schiffman, R.L. (1962), "General Solution of Stresses and Displacements in Layered Elastic Systems", Proceeding of International Conference on the Structural Design of Asphalt Pavement, University of Michigan, Ann Arbor, Michigan, USA.

14. Schofield, A.N., and Wroth, C.P., (1968), Critical State Soil Mechanics, McGraw Hill, London.

15. Yandell, W.O. (1971), "Prediction of the Behavior of Elastoplastic Roads During Repeated Rolling Using the Mechano-Lattice Analogy", Highway Research Record, Highway Research Board, No. 374, pp. 29-41.

Chapter 5

Evaluation of Hot Mix Asphalt Mixtures With Replacement of Aggregates by Reclaimed Asphalt Pavement (RAP) Material

[1]O. Reyes-Ortiz, [1]E. Berardinelli, [2]A.E. Alvarez, [2]J.S. Carvajal-Muñoz and [3]L.G. Fuentes

[1]Nueva Granada Military University, Department of Civil Engineering, Carrera 11 101-80, Bogotá, Colombia

[2]University of Magdalena, Program of Civil Engineering, Carrera 32 No 22-08, Santa Marta, Colombia

[3]Universidad del Norte. Department of Civil and Environmental Engineering, Km 5 Vía Puerto Colombia, Barranquilla, Colombia

ABSTRACT

Due to economical reasons and the need for environmental conservatism, there has been an increasing shift towards the use of reclaimed asphalt pavement (RAP) materials in the pavement construction industry. This is a relatively new concept in Colombia with scarce local documented literature, particularly on the mechanical properties of the hot mix asphalt (HMA) mixtures modified by addition of RAP and the corresponding mixture design procedures. This research focused on evaluating the effects of partial- and total-replacements of aggregates by RAP on the mechanical response of dense-graded HMA mixtures. Corresponding results suggest that the highest indirect tensile strength and resilient modulus values in both dry- and wet-condition were obtained for the HMA mixtures produced with 100% replacement of granular material by RAP.

INTRODUCTION

Recycling of asphalt pavements is a technology developed to rehabilitate and/or replace pavement structures suffering from permanent deformation and evident structural damage [1]. In this context, according to [2], the reclaimed asphalt pavement (RAP) is one of the most recycled materials in the world. The first data documented on the use of RAP for the construction of new roads date back to 1915 [3]. However, the actual development and rise of RAP usage occurred in the 1970's during the oil crisis, when the cost of the asphalt binder (or asphalt) as well as the aggregate shortages where high near the construction sites [4]. Later, in 1997, with the Kyoto Protocol adaptation by parties and implementation in 2005, recycling received major attention and broader application in the road construction industry [5].

Several authors state that diverse methods for recycling of asphalt pavements are suitable including: hotrecycling in plant, hot-recycling "in situ", cold-recycling "in situ", and others [6, 1, and 7]. Nevertheless, hot recycling is one of the most widely techniques used nowadays, where virgin materials and RAP are combined in different proportions and sizes [8]. Studies in Europe and the United States have concluded that over 80% of the recycled material is reused in the construction of roads, but regulations are still strict allowing inclusion of RAP in proportions ranging between 5 and 50% for production of new hot mix asphalt (HMA) mixtures [9]. Studies performed to determine the response of HMA mixtures with RAP replacements between 0 and 40% and fabricated with different asphalts, have shown the low moisture damage susceptibility of the new HMA mixtures (i.e., based on retained tensile strength (RTS) values above 95%; Superpave criteria-ASTM D4867). Similarly, it was found that the resilient modulus values increase regardless of the tests temperature (-18, 0, 25, and 32 °C), type of asphalt (PG-46-40, PG-52-34, and PG-58-28), and addition of RAP (15, 30, and 40%) [10].

According to [2], the incorporation of 40% RAP in HMA mixtures created no modification on the mixture properties. Conversely, when values higher than 40% were included, the mixture properties changed drastically. In general, when higher percentages of RAP were used, evident reductions on the relative energy loss—computed based on the load-displacement curve determined for the indirect tensile test—were reported with possible appearance of premature distresses. The latter can be related to possible moisture damage that may affect the mechanical response (i.e., permanent deformation- and fatigue-response) and mixture performance. Recent researches [1, 8, 11, 12], have established that RAP replacement at proportions above 50% are feasible to produce new HMA mixtures, obtaining satisfactory results in the mechanical properties. Similar fatigue curves were determined for HMA mixtures fabricated with low penetration asphalt (13/22) and HMA mixtures with 60% RAP replacement. Likewise, the susceptibility to moisture damage was low (RTS values close to 95%).

In addition, the HMA mixtures with RAP replacement increased in 50% the indirect tensile strength (ITS) as compared to that of the HMA mixtures fabricated with virgin materials. The energy dissipated during the ITS test also increased by 100% in the HMA mixtures with RAP replacement. Olard et al. (2008) [13] assessed HMA mixtures with high recycling rates (i.e., >50% RAP replacement) for warm- and HMA-mixture production and stated that RAP foster positive environmental impacts, including that it: (i) can be done in an asphalt plant or in-place, (ii) reuses existing materials thus eliminating disposal problems (saving or diminishing land requirements in populated countries), (iii) saves costly materials and in some countries rare, hard to find good aggregates, (iv) can correct both asphalt content and aggregate gradation of an existing HMA mixture, and (v) produces a stable pavement structure at a lower cost than that associated with conventional methods. Based on the positive experiences and outcomes from global use of HMA mixtures with RAP inclusion, it can be inferred that relevant results could be obtained from application of this technology in developing countries, such as Colombia. In this regard, research projects must be conducted and financial support gathered to advance in the development of feasible alternatives tending to be less invasive to the environment and practical in use for constructors and practitioners. Similarly, the same concerns rose by the Kyoto protocol and other global policies with regard to air pollution must be taken into account to minimize risks on human health and ensure environmental quality.

As a contribution to the aforementioned aspects, the main objective of this research is to evaluate the effects of partial- and total-replacements of aggregates by RAP on the mechanical response of dense-graded HMA mixtures specified by the "Instituto de Desarrollo Urbano" (IDU) [14] in Bogota D.C. (Colombia). Through this approach, it is expected to bring to Colombia, and elsewhere, results that may generate new alternatives in the paving industry that led to environmentally sustainable practices and applications. After this introductory section, the paper includes a description of the materials and methods used in this research. Following, results are discussed and the paper concludes with a section of conclusions and recommendations.

MATERIALS AND METHODS

This study was conducted based on the methodology depicted in Fig. 1. The first task of the study comprised the materials' characterization including aggregates, asphalts, and RAP (i.e., gradation and estimation of the asphalt content in the RAP). Subsequently, by applying the Marshall mix-design method, the percentage of asphalt to be added to the new HMA mixtures was determined. The objective of the mix design was to obtain a mixture exhibiting balanced conditions in terms of stability, flow, density, and total air voids content. Then, laboratory specimens were produced based on the aggregate gradation specified for the dense-graded md20 HMA mixture [15] and using two penetration asphalts. These specimens were compacted at 75 blows per face using the Marshall compactor. The study concluded by conducting the mechanical characterization of the HMA mixtures through the indirect tensile test and resilient modulus test

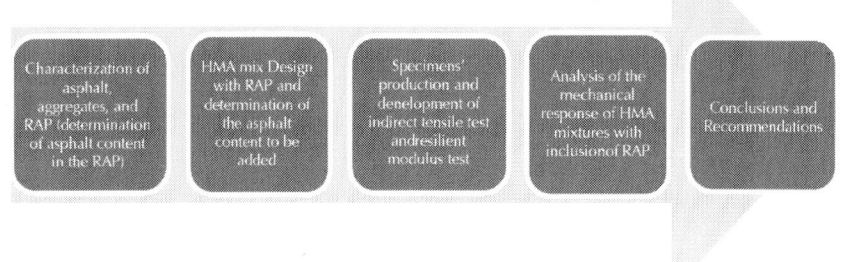

Figure 1: Research methodology conducted for the use of RAP in asphalt mixes.

Table 1: Asphalt characteristics

"Test	Standard	Asphalt 1 (60/70)	Asphalt 2 (80/100)
Penetration (1/10 mm)	ASTM D 5-97	63	89
Ductility (cm)	ASTM D 113-99	120	133
Viscosity (Poises)	ASTM D 2170-95	1500	1250
Softening point (°C)	ASTM D 36-95	47	
Flame and ignition point (°C)	ASTM D 3143-98	235 and 245	220 and 225

Table 1 presents the results on the characterization of the two penetration asphalts used, which corresponded to a 60/70 (1/10 mm) penetration asphalt (or 60/70 asphalt) and an 80/100 (1/10 mm) penetration asphalt (or 80/100 asphalt). Figure 2 shows the gradation curve for the dense-graded md20 HMA mixture from IDU. The aggregate gradation used corresponds to that obtained using the mean values of the specified aggregate gradation band.

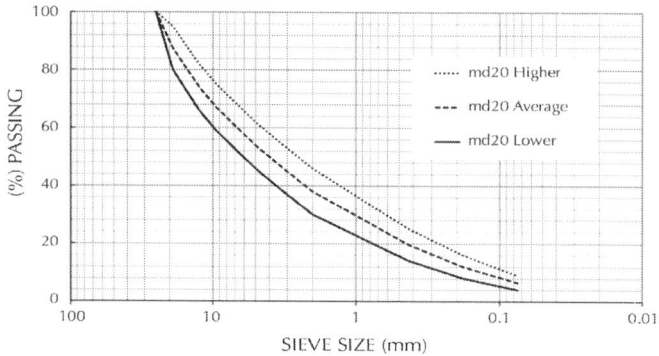

Figure 2: Selected aggregate gradation curve and specified aggregate gradation band for the md20 HMA mixture.

Table 2 presents the asphalt content of RAP evaluated by means of the centrifuge test [16]. This evaluation led to an overall value of asphalt content for all the RAP fractions used. This RAP came from dense-graded HMA mixtures used in Bogota D.C. and its age ranged between 8 and 10 years. Since the RAP gradation was diverse in the stockpiles, the RAP gradation used for the new mixes was produced in the laboratory (i.e., sieving, weighting and mixing; mechanical fractioning of the RAP material was not conducted) to reproduce in all cases—based on different replacements of granular materials by RAP as next indicated—the gradation of the md20 HMA mixture (Fig. 2). The replacements of granular materials by RAP were performed in four manners: (i) full replacement of the aggregates by RAP, (ii) replacement of the aggregate fractions passing the # 4 sieve and retained on the # 10 sieve (i.e., 15% RAP replacement), (iii) replacement of the aggregate fraction passing the # 10 sieve and retained on the # 40 sieve (i.e., 20% RAP replacement), and (iv) replacement of the aggregate fraction passing the # 4 sieve and retained on the # 40 sieve (i.e., 35% RAP replacement). The corresponding HMA mixtures produced are subsequently termed as (i) 100% RAP, (ii) # 4-10 sieves, (iii) # 10-40 sieves, and (iv) # 4-40 sieves. These replacements were selected based on the main size fractions encountered in the RAP material. In addition, future research will allow the comparison of the aforementioned mixtures and a mixture fabricated with 0% replacement of granular material by RAP.

Table 2: Asphalt content of RAP

Specimen weight	Specimen 1	Specimen 2	Specimen 3	Specimen 4	Specimen 5	Mean value
Initial weight (g)	1200.03	1201.01	1200.13	1202.23	1200.52	-
Final weight (g)	1126.86	1127.32	1125.98	1129.2	1123.8	-
Asphalt content (%)	6.10	6.14	6.18	6.07	639	6.18

To determine the percentage of asphalt to be added to the new HMA mixtures produced with RAP, Marshall compacted specimens with addition of 2, 3, 4, and 5% of asphalt were prepared. Fig. 3 shows typical results of stability, flow, and density of the corresponding specimens determined for the HMA mixture fabricated with the # 10-40 sieves RAP replacement. From the Marshall test results (i.e., stability, flow, density, and total air voids content); the percentage of neat asphalt to be added was determined as 3% for all mixes.

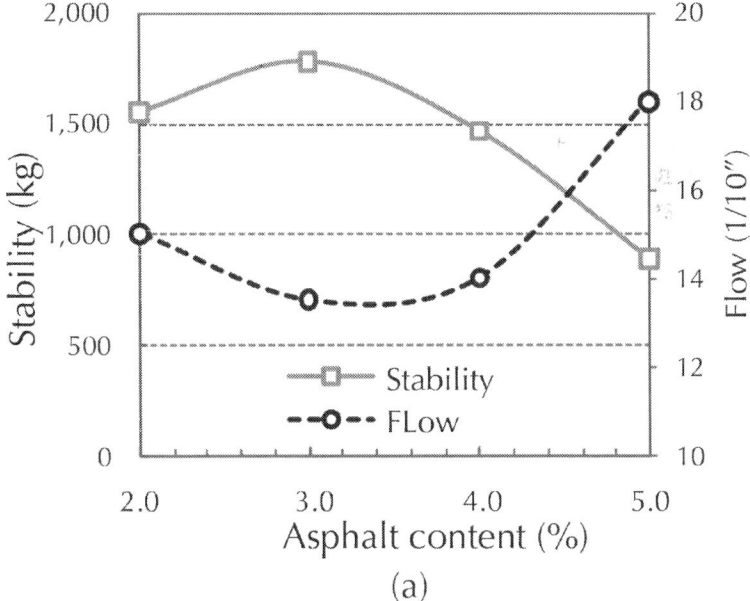

(a)

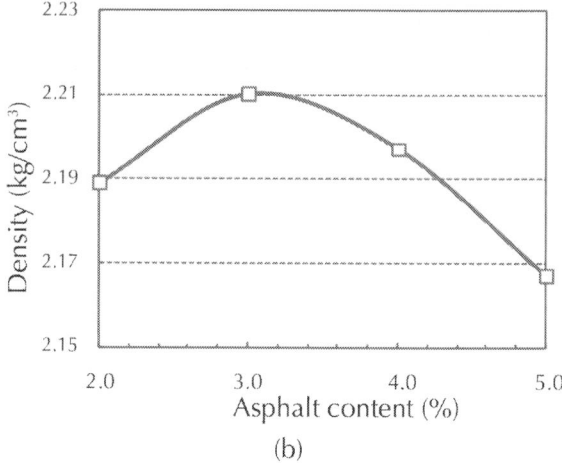

(b)

Figure 3: Stability and flow (a) and density (b) for HMA mixtures fabricated with RAP inclusion.

The mechanical response of the HMA mixtures was then assessed based on the indirect tensile test and resilient modulus test conducted on specimens in dry- and wet-condition. The wet conditioning process corresponded to immersion for 68 to 72 hours in water at $60 \pm 1°C$ in accordance with the European standard EN 12697 [16]. Therefore, the mechanical response parameters evaluated corresponded to the ITS, RTS (ratio of the ITS in wet condition to the ITS in dry condition), and resilient modulus

RESULTS AND DISCUSSION

This section presents the mechanical characterization results for the HMA mixtures analyzed. These results include values of ITS—for specimens tested at both dry and wet conditions—, RTS, and resilient modulus for the specimens fabricated with the dense-graded md20 gradation, two asphalt types, and different RAP replacement proportions. Fig. 4 shows the ITS values in dry condition for the specimens with different RAP replacements. Regardless of the asphalt type, the lowest ITS corresponds to the HMA mixture fabricated with the # 10-40 sieves RAP replacement, while the higher strength was found for the replacement of 100% RAP. The differences reported between the HMA mixtures fabricated with the # 10-40 sieves RAP replacement and the other RAP replacements are substantial and denote that the replacement of the #10-40 sieves aggregate fraction can be critical to ensure the HMA mixture response. Furthermore, as theoretically expected given the asphalt penetration value, the HMA mixtures fabricated with the 60/70 asphalt, regardless of the RAP replacement applied, exhibited a higher ITS as compared to those fabricated using the 80/100 asphalt.

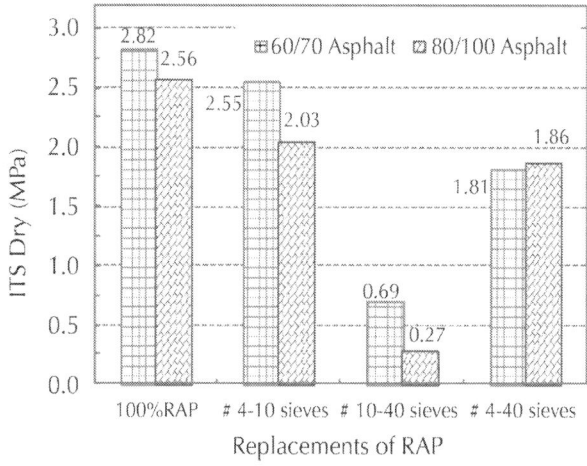

Figure 4: Values of indirect tensile strength (ITS) for specimens tested in dry condition.

Fig. 5 shows the ITS results for specimens tested in wet condition. A similar order of magnitude is reported for the replacements of granular material by 100% RAP, # 4-10 sieves RAP, and # 4-40 sieves RAP, although as previously discussed for the ITS values in dry condition, the specimens produced with 100% RAP replacement exhibited the highest strength values. However, the specimens fabricated with the # 10-40 sieves RAP replacement exhibited the lowest strength, regardless of the asphalt type used. These results are coincident with those previously discussed for the HMA mixtures assessed in dry condition and consistently suggest that the replacement of the #10-40 sieves aggregate fraction can be critical to ensure high values of strength for the dense-graded md20 mixtures analyzed.

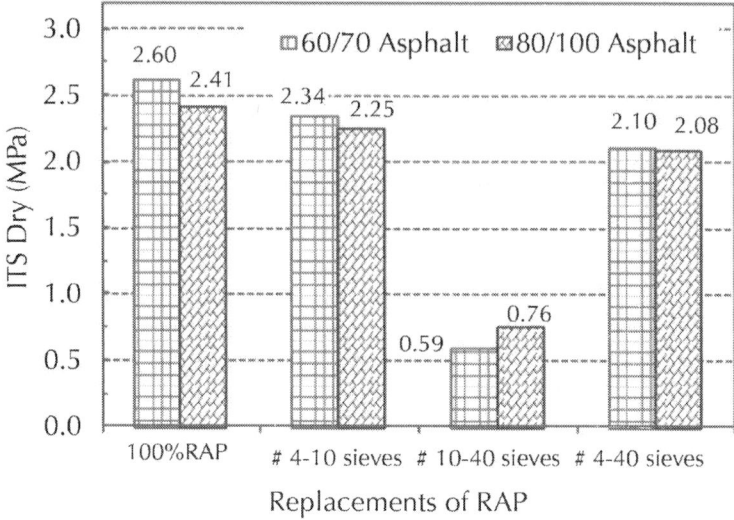

Figure 5: Values of indirect tensile strength (ITS) for specimens tested in wet condition.

The values of RTS computed for the HMA mixtures produced with different replacements of granular material by RAP are presented in Fig. 6. These data indicate low moisture damage susceptibility for the mixtures analyzed based on consistent values of RTS higher than 80%, which is the lower acceptance limit established in the ASTM D4867 standard [15]. The 80/100 asphalt showed to be less susceptible to moisture damage than the 60/70 asphalt. In addition, some RTS values exceeded 100%, showing little effects of the water conditioning process applied in these mixtures produced with RAP replacement. In particular, the RTS values computed for the HMA mixture fabricated with the # 10-40 sieves RAP replacement are comparable to those reported for the other HMA mixtures evaluated. This result indicates that although the tensile strength was low for the HMA mixtures fabricated with the # 10-40 sieves RAP replacement, their moisture susceptibility—as evaluated based on the RTS—is comparable to that obtained when different RAP replacements were tested. This result implies that the #10-40 sieve RAP fraction can be critical to ensure the HMA mixture strength in the mixtures evaluated, although this fraction of added RAP did not trigger substantial modifications in the moisture damage susceptibility of the mixture.

The effect of the #10-40 sieve RAP fractions on the mixture strength can be related to the following aspects:

- Lack of friction between the RAP aggregates due to presence of the former asphalt binder and
- The existence of free asphalt binder in the mixture due to the lack of air voids in the RAP fraction (i.e., inadequate size particle distribution in the RAP fraction).

The moisture damage response can be due to the relatively low proportion (i.e., 20% in the mixture gradation) of the added RAP fraction in the overall mixture composition and mainly to the proper resistance to moisture damage of the asphalt-RAP systems formed in the HMA mixtures evaluated in this research. This aspect is inferred from the RTS values (i.e., higher than 90%) obtained for the HMA mixture fabricated with 100% RAP.

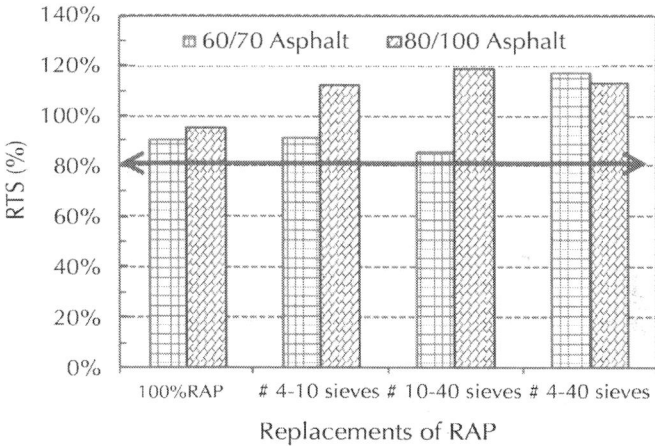

Figure 6: Values of retained tensile strength (RTS).

Fig. 7 and Fig. 8 show the resilient modulus values for specimens produced, respectively, with the 60/70 asphalt and 80/100 asphalt. These specimens were tested in dry condition. As depicted, the HMA mixtures produced with 100% RAP show the highest stiffness values, regardless the test frequency and asphalt type. On the other hand, the HMA mixture produced with the # 10-40 sieves RAP replacement exhibits the lowest modulus values. Nevertheless, when comparing the mixtures fabricated with partial replacement of aggregates by RAP, higher values are reported for the HMA mixture fabricated with the #4-10 sieves RAP replacement.

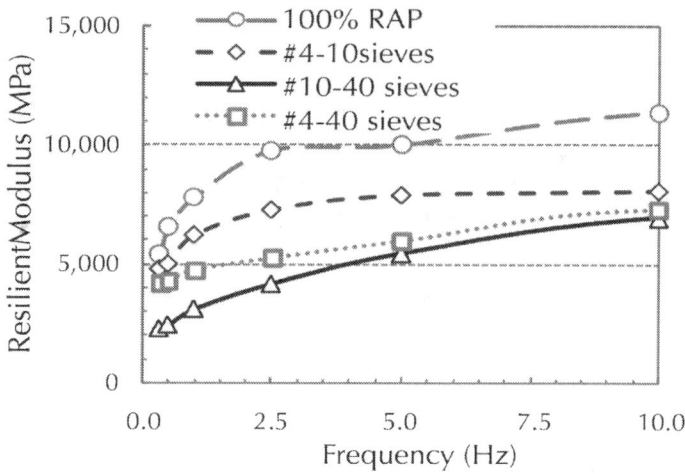

Figure 7: Values of resilient modulus for specimens fabricated with 60/70 asphalt and tested in dry condition.

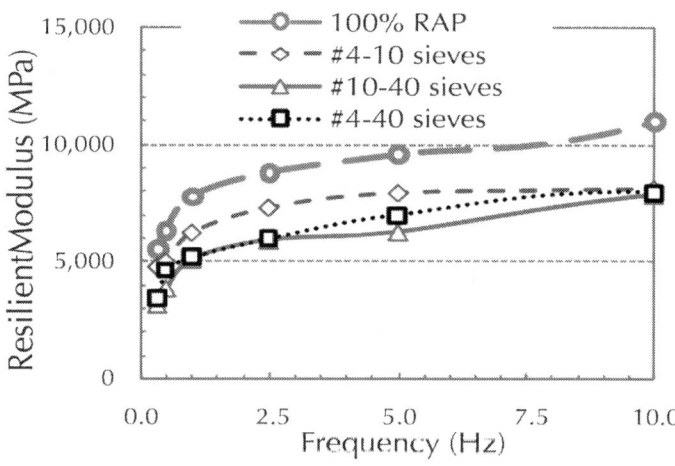

Figure 8: Values of resilient modulus for specimens fabricated with 80/100 asphalt and tested in dry condition.

Fig. 9 shows the resilient modulus values for moisture conditioned specimens and produced by using different replacements of granular material by RAP and the 60/70 asphalt. It is observed that higher modulus values are again obtained for the HMA mixtures fabricated with 100% RAP. The lower moduli values are again found for the specimens produced with the # 10-40 sieves RAP replacement. Comparison of the modulus values obtained for the HMA mixtures evaluated in dry condition (Fig. 7) and wet condition (Fig. 9) provides evidence of the negative effect of water on the mixture stiffness (i.e., moisture damage effect). Additional research is still required to further quantify the differences in moisture damage susceptibility of mixtures fabricated using different proportions of RAP (i.e., different proportions of aged asphalt binder). The HMA mixture more affected by moisture damage corresponded to that fabricated with the # 10-40 sieves RAP replacement. A different conclusion was previously stated based on the analysis of the RTS values, providing evidence of restrictions in the RTS to rank and differentiate the moisture damage susceptibility of HMA mixtures. These findings are coincident with those reported in previous literature [17].

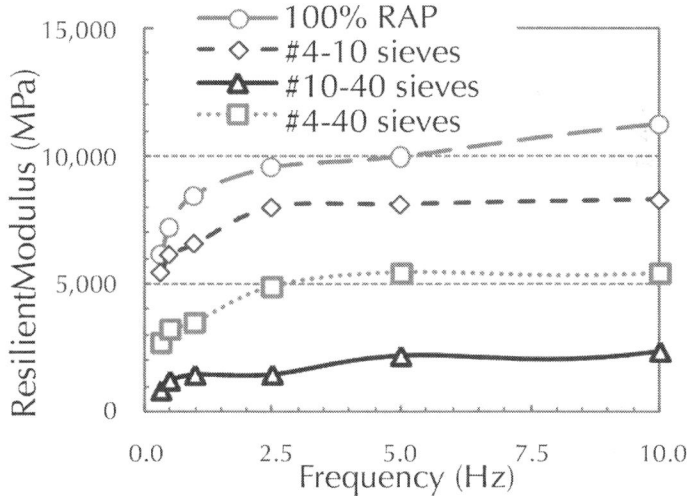

Figure 9: Values of resilient modulus for specimens fabricated with 60/70 asphalt and subjected to moisture conditioning.

Similar conclusions can be stated based on the Fig. 10, which shows the values of resilient modulus for specimens fabricated with the 80/100 asphalt and tested after subjecting them to moisture conditioning.

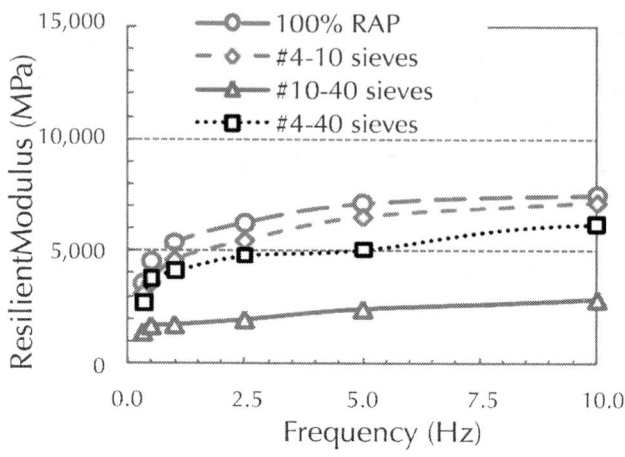

Figure 10: Values of resilient modulus for specimens fabricated with 80/100 asphalt and subjected to moisture conditioning.

CONCLUSIONS

This paper presents an evaluation of the effect of conducting partial- and total-replacements of aggregates by RAP to fabricate dense-graded HMA mixtures. Hence, the research focused on evaluating the mechanical response of the HMA mixtures produced with inclusion of RAP. This evaluation was conducted in terms of the indirect tensile strength (ITS) in dry- and wet-condition, retained tensile strength (RTS), and resilient modulus (evaluated in dry- and wet-condition) of HMA mixtures fabricated with four diverse RAP replacements. Based on the laboratory results and corresponding analysis conducted, the next conclusions are offered:

- The highest ITS and resilient modulus values in both dry- and wet-conditions were obtained for laboratory specimens fabricated with 100% replacement of granular material by RAP, independently of the asphalt type used in the HMA mixture.

These results show the potential effect of the aged asphalt (stiffer asphalt binder as compared to that used to fabricate the new mixtures) contained in the RAP on the mechanical response of the recycled mixtures.

- Replacement by RAP of the granular material in the fraction passing the # 10 sieve and retained on the # 40 sieve, exhibited the most unfavorable conditions, regardless of the testing condition (i.e., dry or wet), asphalt type, and testing carried out (i.e., ITS and resilient moduli). Therefore, replacement by RAP of the aggregate fraction aforementioned was identified as critical for the mechanical response of the HMA mixtures evaluated. This aspect can be related to the lack of friction between the RAP aggregates and/or the existence of free asphalt binder in the mixture due to the lack of air voids in the RAP fraction.

- For all the HMA mixtures tested with various replacements of granular material by RAP, a low susceptibility to moisture damage was reported as evaluated in terms of the RTS values. In fact, all the RTS values reported were above 80% and, in certain cases, values above 100% were reported. However, different conclusions were obtained in terms of the resilient modulus values, which suggest the need for additional research to better evaluate the moisture susceptibility of HMA mixtures fabricated with inclusion of RAP.

- Higher ITS values (in both dry- and wet-condition) were reported for HMA mixtures fabricated with the 60/70 asphalt as compared to those determined for the HMA mixtures manufactured with the 80/100 asphalt. However, for the RTS, the highest values were associated with the 80/100 asphalt.

- The conclusions before indicated are restricted to the materials combinations evaluated. Additional research is suggested to comprehensively assess more material combinations and validate the corresponding findings through field performance evaluations.

ACKNOWLEDGEMENTS

The authors gratefully thank the support provided by the Nueva Granada Military University (UMNG) and the Geotechnical Engineering

Research Group for the laboratory tests executed. Also, thanks are provided to the Vice research Office for the financial support on the ING-730 project.

REFERENCES

1. Valdés, G., Pérez-Jiménez, F., Miró, R., Martínez, A., & Botella, R. (2011). Experimental study of recycled asphalt mixtures with high percentages of reclaimed asphalt pavement (RAP). Construction and Building Materials, 25(3), 1289 – 1297

2. Chen, J., Wang, C. & Huang, C. (2009). Engineering properties of bituminous mixtures blended with second reclaimed asphalt pavements (R2 AP). Road Materials and Pavement Design, 10, 129 – 149.

3. Taylor, N. (1997). Life expectancy of recycled asphalt paving. Recycling of bituminous pavements (L. E. Wood, Ed.). ASTM STP 662, 3–15.

4. Sullivan, J. (1996). Pavement recycling executive summary and report. Report FHWA-SA-95-060 from the Federal Highway Administration. Washington, D.C.

5. Reyes, O., Pérez, F., Miro, R., & Botella, R. (2009). Proyecto fénix. Mezclas semicalientes. Proceedings del XV Congreso Iberolatinoamericano del Asfalto. Lisboa, Portugal.

6. Decker, D. (1997). State of the practice for use of RAP in hot mix asphalt. Journal of the Association of Asphalt Paving Technologists, 66, 704.

7. Silva, H., Oliviera, J, & Jesus, C. (2012). Are totally recycled hot mix asphalts a sustainable alternative for road paving? Journal Resources, Conservation and Recycling, 60, 38 - 48.

8. Reyes, O., & Camacho, J. (2012). Informe Proyecto ING-730 Estudio del comportamiento de mezclas asfálticas colombianas al adicionarles RAP en diferentes porcentajes y tamaño. Reporte de la Universidad Militar Nueva Granada. Colombia.

9. Mengqi, W., Haifang, W., Muhunthan, B., & Kalehiwot N. (2012). Influence of RAP content on the air void distribution, permeability and moduli of the base layer in recycled asphalt pavements.

Proceedings of the 91st Transportation Research Board Meeting. TRB 2012. Washington, D.C., USA.

10. Sondag, M., Chadbourn, B., & Drescher, A. (2002). Investigation of recycled asphalt pavement (RAP) mixtures. Technical report MN/RC – 2002-15. Department of Civil Engineering, University of Minnesota. USA.

11. Pereira, P., Oliveira, J, & Picado-Santos, L. (2004). Mechanical characterization of hot mix recycled materials. International Journal of Pavement Engineering, 5 (4), 211 - 220.

12. Celauro, C., Bernardo, C., & Gabriele, B. (2010). Production of innovative, recycled and high-performance asphalt for road pavements. Journal Resources, Conservation and Recycling, 54 (6), 337 - 347.

13. Olard, F., Noan, C., Bonneau, D., Dupriet, S., & Alvarez, C. (2008). Very high recycling rate (>50%) in hot mix and warm mix asphalts for sustainable road construction. Proceedings of the 4th Eurasphalt and Eurobitume Congress. Copenhagen, Denmark.

14. Instituto de Desarrollo Urbano -IDU- (2005). Especificaciones técnicas generales de materiales y construcción para proyectos de infraestructura vial y de espacio público de Bogotá. Reporte de IDU, Colombia.

15. ASTM International. (2011). Annual book of ASTM standards. West Conshohocken, PA.

16. European standard (2004). Bituminous mixtures – test methods for hot mix asphalt. EN 12697-12: 2003–Part 12 - Determination of the water sensitivity of bituminous specimens.

17. Caro, S., Masad, E., Bhasin, A., & Little, D.N. (2008). Moisture susceptibility of asphalt mixtures, Part I: mechanisms. International Journal of Pavement Engineering, 9, 81 - 98.

6

Performance Characteristics of Fiber Modified Asphalt Concrete Mixes

Manoj Shukla[1]/Dr. Devesh Tiwari[2]/K. Sitaraman-janeyulu[3]

[1]Senior Scientist, Central Road Research Institute (CRRI), New Delhi, India

[2]Principal Scientist, Central Road Research Institute (CRRI), New Delhi, India

[3]Senior Principal Scientist, Central Road Research Institute (CRRI), New Delhi, India

ABSTRACT

Asphalt binder modification is one of the approaches taken to improve pavement performance. In addition it may also be improved through the addition of fibers to Asphalt mix that enhances material strength and fatigue characteristics while adding ductility. Due to their inherent compatibility with Asphalt concrete and excellent mechanical properties, fibers offer an excellent potential for modification of Asphalt concrete mix. To investigate the behavior of Fiber Modified Asphalt Concrete Mixes (FMACM), a preliminary study has been done to determine the feasibility of modifying the behavior of a Asphalt Concrete (AC) mixture through the use of Glass fiber and Polyester fiber. The purpose of this study was to identify and understand the factor that is responsible for improving the behavior of FMACM. Asphalt concrete samples were prepared and tested in the laboratory to evaluate the various mixture characteristics. The conclusions drawn from the study on testing of fiber-modified mixes are that fiber modified Asphalt mixtures have shown increased stiffness and resistance to permanent deformation. Fatigue characteristics of the mixtures were also improved. Fibers used in the study were of high tensile strength therefore test results of FMACM have shown higher indirect tensile strength and improved skid resistance for paving applications.

INTRODUCTION

Modification of Asphalt Concrete Mix can be done in many ways such as modification of binder through polymers, crumb rubber etc. or reinforcing of AC mixes. Fiber reinforcement was used as a crack barrier rather than a reinforcing element whose function is to carry the tensile loads as well as to prevent the formation and propagation of cracks. Thin fibers exhibit much higher rupture stresses than the same in thicker form. The strengthening mechanism operates because the glass fiber carries the applied stress over a high tenacity cross-section and distributes it to the weaker material as a shearing stress over a relatively long periphery. Synthetic fibers provide a different mode for extending the life of overlays. They re-distribute the strain in an overlay immediately above a crack reflecting upward. Glass fibers are more effective against bending and thermal contraction displacement than against shear displacement. In the present study laboratory test involved are beam fatigue test, indirect tensile tests, repeated load creep and permanent deformation tests, besides all other common tests such as density, Marshall Stability, moisture susceptibility, and others. This study investigates on the characteristics and properties of FMACM, which may have the benefit for improving the performance of pavement.

Literature Review on Fibers Modified Asphalt Mixes

The cellulose fibers are currently being used extensively in porous asphalt friction course, with the primary purpose of increasing asphalt content without any drain down problem (Busching et al. 1968). The fiber reinforces the binder system, thus causing an increase in the viscosity of the system. The resulting mix has greater stability and possibly higher resistance to fatigue cracking than similar mixes without addition of fibers. The mineral fibers are typically heavier than the cellulose fibers and therefore is used at a higher content by weight in mixture so that sufficient volume of fibers is available.

Serfass et al. (1966) examined the effects of fiber modified asphalt utilizing asbestos, rock wool, glass wool, and cellulose fibers and concluded that adding fibers enables developing mixtures rich in asphalt and therefore displaying high resistance to moisture, aging, fatigue, and cracking. Simpson et al. (1984) conducted a study of modified asphalt mixtures in Somerset, Kentucky. Polypropylene, polyester fibers and Crumb Rubbers were used to modify the asphalt binder. Two proprietary blends of modified binder were also evaluated. An unmodified mixture was used as a control.

Mixtures containing polypropylene fibers were found to have higher tensile strengths and resistance to cracking. Rutting potential as measured by repeated load deformation testing was found to decrease in modified samples. The suitability of a mineral fiber for use in road asphalt is thus dependent on two criteria i.e. average fiber diameter and average fiber length. Though both fibers are not similar in shape (cross section) but the length are almost similar. Therefore, the laboratory performances have been compared with respect to different tests.

The use of Glass fiber showed consistent results (Abdelaziz et al. 2005) and it was found that the addition of fiber does affect the properties of asphalt mixes, by decreasing its stability and an increase in the flow value as well as the voids in the mix. The results indicated that the fiber has the potential to resist structural distress that occur in road pavement as result of increased traffic loading, thus improving fatigue life by increasing the resistance to cracking and permanent deformation. On the whole, the results showed that the addition of glass fiber will be beneficial in improving some of the main properties of the flexible pavement. Praveen et al. (2009) has reported that the indirect tensile strength increases with an increase in fibers content in the mix. The tensile strength ratio and retained stability increase with an increase in fibers content, indicating improved moisture resistance. Permanent deformation decreases with an increase in fibers content, indicating a decrease in rutting potential. Decoene,Y (1990) have concluded that use of cellulose fibers is suitable for porous Asphalt mixes. Jiang et.al (1993) and Maurer, et al. (1989) emphasised on use of fibers to retard reflective cracking in pavements.

ASPHALT CONCRETE MIXTURE CHARACTERIZATION

Material

For the laboratory study, Delhi Quartzite aggregate of sizes 20 mm, 10 mm, stone dust and lime were used and tested as per IS Standards. Asphalt of Viscosity Grade-30 (VG-30) was used and tested as per Bureau of Indian Standards (BIS) (1978). The results of tests performed on asphalt for the study are shown in Table 1 and basic tests on aggregates performed during the study are shown in Table 2 as per BIS (1978) and limits as per the specifications of Ministry of Road Transport and Highways (MoRTH) (2001). The asphalt mix design was done on a control gradation of Asphalt Concrete (AC) mixture as per MoRTH (2001) which is shown in Fig-1. The MoRTH specification also states that either of one test is needed i.e. Aggregate Impact Value test or Los Angeles Abrasion test, therefore in the present research only Aggregate Impact Value test has been done.

Table 1: Test results of asphalt.

Sr. no.	Test description	Results
1	Penetration at 25 °C, 100 g, 5s.	7 mm
2	Ductility at 27 °C	98 cm
3	Specific Gravity	1.01 -
4	Softening Point	49°C
5	Flash Point	285°C
	Fire Point	296°C
6	Viscosity(Pa s) at 135 °C	5.4

Table 2: Test results of aggregates

S.No.	Test description	Results	Limits
1	Combined Flakiness & Elongation Index	28 %	< 30
2	Specific Gravity		
	20 nun	2.669	-
	10 nun	2.665	-
	Stone Dust	2.679	-
	Lime	2.800	-
3	Water Absorption	1.2 %	≤2
4	Aggregate Impact Value	16.5 %	≤30
5	Stripping Value	5 %	≤5

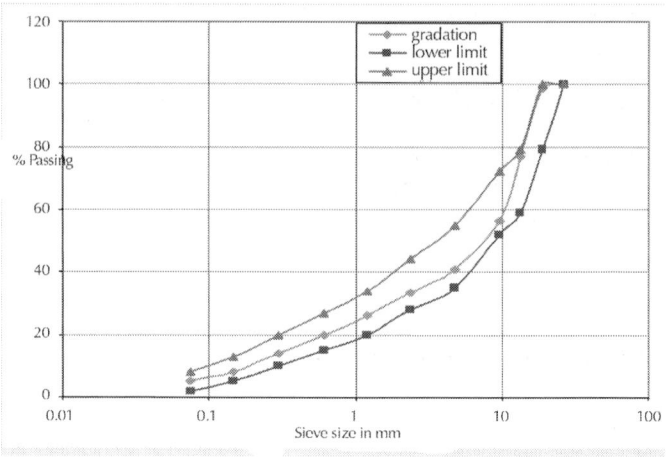

Figure 1: Aggregate gradation of asphalt concrete mix

Asphalt Concrete Mix Design

Marshall Method as per Asphalt Institute Manual (MS 2, 1997) of mix design was done for determination of optimum asphalt content (OAC) and found to be 5.3 % by weight of aggregate. The properties of the Marshall design and limits for AC as per MoRTH specifications are given in Table 3.

Table 3: Asphalt concrete mix design values

Sr. No.	Parameters	Results	Specified Limit
1.	Stability (kN)	11.65	9 minimum
2.	Flow (mm)	3.1	2 - 4
3.	Density (g/cm3)	2.362	-
3.	Air Voids, %	5	3 - 6
4.	Voids Filled with Asphalt (VFB). %	72	65 - 75
5.	Retained Stability (%)	84	75 minimum

Reinforcement of Asphalt Mix Using Glass Fiber and Polyester Fiber

Marshall Mix design using glass fiber was carried out in the laboratory through the following steps. The four major steps used for modified mix design are briefly discussed here.

1. Materials (fiber) selection.

2. Mix procedure.

3. Selection of optimum glass fiber content to Binder content ratio (Fc/AC).

4. Various tests performed on optimum Fc/AC ratio.

Material (Fiber) Selection

The glass fiber and Polyester fiber which were selected for the present study has melting point much higher than required by the asphalt mix during mixing and lying. The test results of glass fiber and Polyester fiber properties are given in Table 4.

Table 4: Properties of glass and polyester fibers used

Test	Glass Fiber	Polyester Fiber
Color (visual)	White	White
Shape	Rectangular	Circular
Fiber width (mm)	1.0	0.002
Fiber Cut Length (mm)	10	8
Tensile Strength (MPa)	1400	9×104
Modulus of Elasticity (MPa)	70000	8×106
Melting Point (°C)	> 300	>250
Specific Gravity (g/cm3)	2.52	1.34
Moisture - (%)	< 0.20	< 0.5
Loss on Ignition - (%)	< 0.25	< 0.5
Non-Fibrous Material - (%)	< 1.00	< 2.5

Mixing Fibers with Asphalt Mix

As the glass and polyester fibers were not uniformly dispersing in the binder therefore experiment was performed to calculate the melting point of the fibers and after experimentations it was seen that the fibers had high melting point i.e. > 300 °C and >250°C for glass and polyester fibers respectively. The mixing temperature of asphalt is around 150 °C - 160 °C so the fibers were not blended with the binder but are added to the mix. An automatic mixer was used to mix the fibers for five minutes known as dry mixing.

Selection of Optimum Fiber to Binder Ratio (Fc/AC)

The volumetric analysis of various parameters in asphalt mix design at varying asphalt content and doses of fibers were performed to arrive at glass fiber/Polyester fiber to binder ratio as the same was adopted for finding out Optimum Binder Content (OBC). Marshall Samples were made using glass and polyesters fibers and also calculated for various tests for volumetric analysis. Air voids were kept around 5- 5.5% with voids filled with asphalt in the range of 68 -75 %. With glass fibers OBC was found to be 5.32 % with glass fiber content of 0.15 percent and for polyester fiber, OBC was found to be 5.35% with fiber content of 0.20 percent.

TEST RESULTS AND ANALYSIS

A number of tests were done to characterize the mechanical properties of AC mixtures. The tests used in the study concerned resilient modulus, repeated load deformation, flexural beam fatigue, indirect tensile and skid resistance. Each of these tests is described in brief with test results of samples made with and without fibers.

Marshall Stability Test

The Marshall Stability test was done as per ASTM D 1559 (1989) and AASHTO T245 (2008). Stability of an asphalt mix depends on internal friction and cohesion. Fibers provide better bonding in the mix. The test results of various mixes with and without fibers are given in Table 5. The retained Marshall stability as per MS 14 (Appendix H) is given in Table 6.

Indirect Tensile Strength

This test measures the splitting tensile strength of asphalt mixes. The indirect tensile strength (ITS) test which is tensile strain at failure is more useful in predicting cracking potential. The test was done as per ASTM D 7369 (2011). The test results of various mixes with and without fibers are given in Table 5. The retained tensile strength (tensile strength ratio) as per AASHTO T 283 (2007) is also given in Table 6.

Table 5: Marshall Stability and ITS

Tests	Plain AC mix	AC Mix With glass fibers	With polyester fiber
Average Stability, KN	11.60	12.85	13.10
ITS (MPa)	89	111	108

Table 6: Retained stability and tensile strength ratio

Tests	Plain AC mix	AC Mix With glass fibers	AC Mix With polyester fiber
Retained Stability,%	84	89	84
Tensile strength ratio,%	89	93	94

Diametrical Resilient Modulus

The resilient modulus of a mixture is a relative measure of mixture stiffness. The test procedure used in this study was ASTM D 4123-82(1995). The test was performed by applying five haversine load pulses of 2000N on one diametric axis and measuring the deformation on the perpendicular axis. Because of the non-destructive nature of resilient modulus testing, the same samples were tested after rotating through

90° and the average of the two values was taken. Three samples were tested at each temperature. Testing was done at two temperatures 25 °C and 35 °C. The test results are given in Table 7.

Table 7: Resilient modulus of asphalt concrete mixes

Parameter	Plain Mix	Mix Reinforced	
		With Glass Fiber	With Polyester Fiber
Average Resilient Modulus at 25°C (MPa)	3109	4287	4226
Average Resilient Modulus at 35°C (MPa)	1077	1282	1446

Repeated Load Deformation

Repeated load deformation tests (Hafeez Imran et.al 2011 & Vos, K.B. 2002) are useful in comparing the rutting susceptibility of different mixtures. In Universal Testing Machine UTM-5P, strain test applies a repeated load of defined magnitude and cyclic duration to specimens. For the current study, a static conditioning stress and conditioning time was applied prior to commencement of actual test.

Conditioning time and conditioning stress were kept 100 s and 10 KPa respectively. In most procedures, the point at which tertiary flow begins is of greatest interest. In this study test was done using UTM-5P. The load pulses required to reach 25,000 micro strain (2.5%) were used as an additional indicator, designated as the cycles to failure or 3600 s of loading whichever is earlier. Test results of accumulated strain on mix samples at 40° C and 60° C are shown in Table 8. Test results indicated that accumulated creep strain in FMACM Mix is lower than plain asphalt mix. Fig 2 shows the pattern of creep stiffness and accumulated strain in repeated load test.

Table 8: Accumulated strain of asphalt concrete mix at 40°C and 60°C

Parameter	Plain Mix at 40°C	Plain Mix at 60°C	AC With Glass Fiber at 40°C	AC With Glass Fiber at 60 °C	AC With Polyester Fiber at 40°C	AC With Polyester Fiber at 60°C
Accumulated Strain %	0.92	2.5	0.75	1.62	0.72	1.74

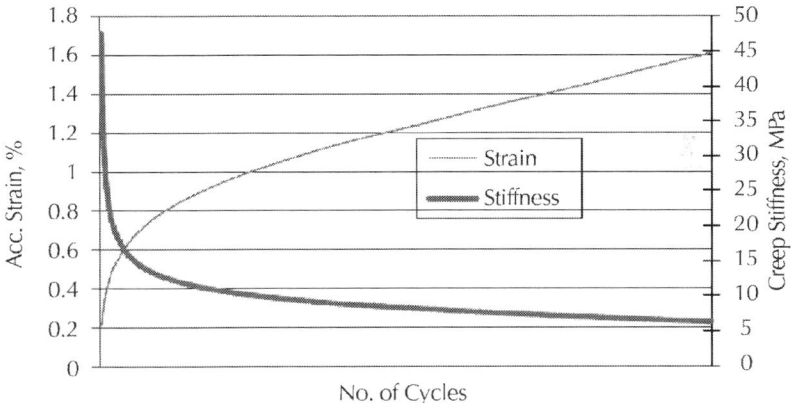

Figure 2: Strain vs. stiffness for creep test for mix with glass fiber at 60°C.

Flexural Beam Fatigue Test

The results from a beam fatigue test can be used to provide an estimation of the number of wheel loads that can be carried by a pavement before fatigue cracking appears. In this study, the fatigue properties of the FMACM and unmodified asphalt mixtures were measured and compared. Each of the asphalt mixtures used for the beam fatigue test was prepared using standard laboratory procedure. Beam fatigue samples were prepared and tested in accordance with ASTM D 7460 (2010) at 25 °C.

Fatigue failure is defined as the point where 50 % of the initial stiffness of the beam specimen remains and the number of cycles are known as fatigue life cycles. The test jig set-up for fatigue test is shown in Photo 1. The test results for flexure stiffness of flexure beam fatigue test at constant strain mode for 300 & 500 micro-strains for plain AC mixes and FMACM and number of fatigue life cycles are shown in Table 9. FMACM has sustained higher number of cycles to reach level of 50 % of initial stiffness compared to plain mix as shown in Figure 3.

Table 9: Comparisons for Initial Stiffness and Number of Fatigue Life Cycles of Mixes at 25°C

TYPE OF MIX		Plain AC	AC mix With glass fiber	AC mix With Polyester fiber	Plain AC Mix	AC mix With glass fiber	AC mix With Polyester fiber
		Initial Stiffness (MPa)			Number of Fatigue Life Cycles		
Micro - Strain	300 μ	1265	1735	1564	152088	196642	212860
	500 μ	1089	1397	1247	41064	52560	59762

Photo 1: Flexure Beam Fatigue Test Jig

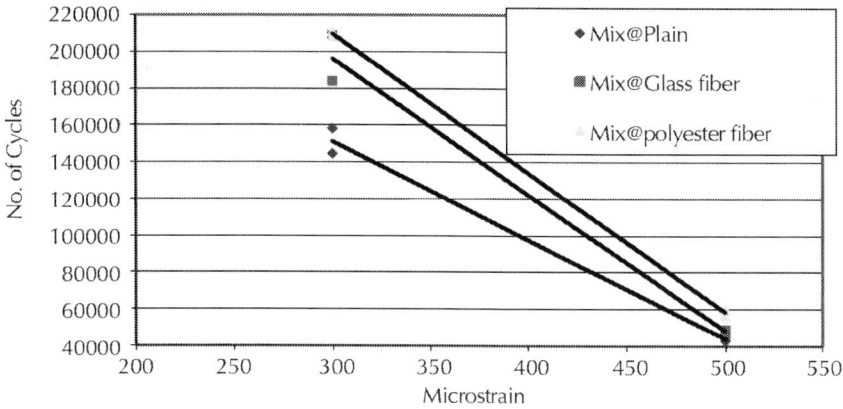

Figure 3: Fatigue life cycle of bituminous mixes

Hamburg Wheel Tracking Test

The Hamburg Wheel-Tracking Device (Sousa, J.B. et.al 1998 & AASHTO T 324, 2011) measures the combined effects of rutting and moisture damage by rolling a steel wheel across the surface of an asphalt concrete slab that is immersed in hot water. The device was developed in the 1970's by Esso A.G. of Hamburg, Germany, based on a similar British device that had a rubber tire. The machine was originally called the Esso Wheel-Tracking Device. This test method covers the determination of rut depth of rectangular specimen (slab) of asphalt mix of size 300 x 150 x 75 mm. Rutting in the specimen occurs due to repetitive action of wheel subjected to standard axle load.

The contact area between wheel and specimen is about 5.45cm² giving a mean normal pressure 566 KPa. The depth of the impression being recorded at the midpoint of its length by means of a rut depth measuring device. The slabs were compacted with static compression machine to achieve compaction level of 5 to 6 percent air voids in the mix. Each sample is loaded at 50 °C for 20000 passes or until 20 mm depression whichever occurs earlier. The total rut depth at 20000 passes for plain AC mix and Fiber modified mix is given in Table 10 and the pattern of rutting is shown in Figure 4.

Table 10: Rutting behavior of various mixes at 50°C

Type of Mix	Plain AC mix	AC mix with glass fiber	With polyester fiber
Total rut depth, mm	14 mm	12mm	11mm

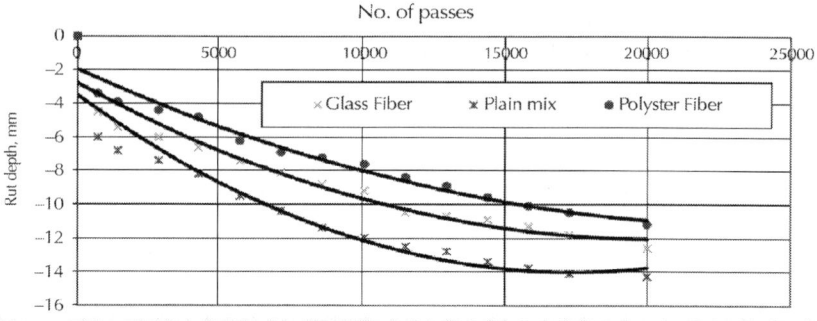

Figure 4: Rut Depth for bituminous mixes

Skid Resistance of Various Mixes

Asphalt Slabs of sizes 350 x 150 x 75 mm made for rutting test were used to measure the skid resistance (at the centre and top of slab) of various mixes using British pendulum skid resistance tester as per ASTM E 303 (1993). During test it was found that skid resistance on asphalt surface improved which may indicate that the tensile reinforcement of fibers to aggregate and binder is strong enough as also shown in Table 11.

Table 11: Coefficient of skid resistance of various mixes

Coefficient of skid resistance	Plain AC mix	AC Mix With glass fiber	AC Mix With polyester fiber
Dry condition	0.81	0.88	0.83
Wet condition	0.72	0.80	0.76

DISCUSSIONS AND CONCLUSIONS

- With the addition of glass fiber and polyester fiber in Asphalt concrete mix, the optimum binder content increased by 0.02% and 0.05 % respectively. Test results indicate that fibers require slightly more bitumen for coating.

- With the addition of glass fiber and polyester fiber in Asphalt concrete mixes, the Marshall stability increases upto 10% and 13% respectively.

- With the addition of glass fiber and polyester fiber in Asphalt concrete mix, the indirect tensile strength increases upto 25% and 21% respectively, whereas tensile strength ratio also increased upto 5% for both type of Fibers, which indicates that cracking in Asphalt concrete mixes may be delayed with addition of fibers.

- With the addition of glass fiber and polyester fiber in Asphalt concrete mix, the average resilient modulus increases upto 38% and 36% respectively at temp of 25°C whereas the increase in resilient modulus is upto 19% and 34% respectively at temperature of 35°C.

- FMACM have shown significantly less Accumulated creep strain i.e. 18% & 22% less at 40°C and 35% & 38% less at 60°C respectively for glass fiber and polyester fiber. Test results indicate the behavior of mix with polyester fiber changes with increase in temperature as the same is susceptible to less strain at 60°C compare to mix with glass fiber.

- FMACM have shown significant improvement 29% and 40% in Fatigue life cycles of the mix at 25°C at control strain mode at strain level of 300μ for glass fiber and polyester fiber respectively whereas improvement of 28% and 46% in Fatigue life cycles of the mix at 25°C at control strain mode at strain level of 500µ for glass fiber and polyester fiber respectively.

- FMACM are less susceptible to rutting, as test results indicated that at 50°C, for 20000 passes, the total rut depth of FMACM is significantly less by 14% & 21% for glass fiber and polyester fiber respectively. However test results indicate that polyester fiber has shown better performance in rutting test but not in Accumulated creep test at 60°C.

- The ability of fiber to increase the tensile strength of Asphalt concrete mixtures at low temperatures have indicated higher modulus and fatigue life cycles. Rutting potential of FMACM at 60°C temperature also indicated better performance for hot climatic condition.

- Glass fiber modified mix has also shown improvement in the skid resistance by 8% and 11% in dry and wet conditions respectively on samples prepared and tested in laboratory whereas there is not much improvements in skid resistance observed in Polyester fiber modified mix.

REFERENCES

1. AASHTO T 245, 2008 "Standard Method of Test for Resistance to Plastic Flow of Asphalt Mixtures Using Marshall Apparatus".

2. AASHTO T 283, 2007 "Standard Method of Test for Resistance of Compacted Asphalt Mixture to Moisture-Induced Damage".

3. AASHTO T 324, 2011 "Standard Method of Test for Hamburg Wheel-Track testing of Compacted Hot Mix Asphalt".

4. Abdelaziz Mahrez, et. Al 2005, "Fatigue and Deformation Properties of Glass Fiber Reinforced Asphalt Mixes", *Journal of the Eastern Asia Society for Transportation Studies,* Vol. 6, 2005 pp. 997 – 1007.

5. ASTM E 303-93, 2008 "Standard Test Method for measuring surface frictional properties using the British Pendulum tester".

6. ASTM D 7460, 2010 "Standard Test Method for determining Fatigue Failure of Compacted Asphalt Subjected to Repeated Flexure Bending".

7.   ASTM D 1559, 1989 "Standard Method of Test for Resistance to Plastic Flow of Asphalt Mixtures Using Marshall Apparatus".

8.   ASTM D 7369 2011 "Standard test Method for determining the Resilient Modulus of Asphalt concrete Mixture by Indirect Tension Test".

9.   Busching H.W. and J.D. Antrim 1968 "Fiber Reinforcement of Asphalt Mixtures." *Journal of the Association of Asphalt Paving Technologists*, Vol. 37, 1968 p 629-656.

10.   Decoene, Y. 1990 "Contribution of cellulose fibers to the performance of porous asphalts", *Transportation Research Record* n 1265, 1990, p. 82-88.

11.   Imran Hafeez & Mumtaz Ahmed Kamal, 2011 "Repeated Load Permanent Deformation Behavior of Mixes With and Without Modified Bitumen", *Mehran University Research Journal of Engineering & Technology*, Volume 30, No. 1, January, 2011.

12.   IS: 1201 to 1220,1978 "Indian Standard Methods for Testing Tar and Asphalt", IS: 1201 to 1220, *Bureau of Indian Standards*.

13.   IS: 2386, 1978 "Indian Standard Methods of Test for Aggregate for Concrete", IS:

14.   2386, *Bureau of Indian Standards*.

15.   Jiang, Yi; Rebecca S. McDaniel., 1993 "Application of Cracking and Seating and Use of Fibers to Control Reflection Cracking" *Transportation Research Record* n 1388, p150-159.

16.   Maurer, Dean A.; Gerald Malasheskie., 1989 "Field performance of fabrics and fibers to retard reflective cracking" *Transportation Research Record No. 1248, 1989, p 13-23*.

17.   "Mix Design Methods for Asphalt Concrete and other Hot Mix Types" 1997, *The Asphalt Institute, Manual (MS-2)*, USA.

18.   Praveen Kumar, et.al. 2009 "Investigation of Fiber Modified Asphalt Mixes", *Journal of Transportation Research Board, Transportation Research Board of the National Academies*, Vol. 2126, 2009 p 91-99.

19.   Serfass, J.P.; J. Samanos. 1966 "Fiber-Modified Asphalt Concrete Characteristics", Applications and Behavior. *Journal of the Association of Asphalt Paving Technologists*, Vol. 65, 1966 p 193-230.

20. Simpson, et. al 1994, "Case study of modified asphalt mixtures: Somerset, Kentucky" *Proceedings of the Third Materials Engineering Conference 804 Oct 1994, ASCE p 88-96.*

21. Sousa J.B. et. al. 1998 "Bridged procedure to determine permanent Deformation in Asphalt Concrete Pavements" *Transportation Research Record 1448*, TRB, National Research Council, Washington DC, pp 25-33.

22. Specifications for Road and Bridge Works, Fourth Revision, 2001 "Physical Requirements of Aggregate for Asphalt Concrete", *Ministry of Road Transport and Highways (2001), Indian Roads Congress*, New Delhi, India.

23. Vos, K.B. et al. 2002 "Feed Back Controlled Repeated Uniaxial Loading Strain Test," Hardware Reference Manual, *Universal Testing Machine (UTM-5P)*, Test No. 20, pp. 1-21, September, 2002.

Chapter 7

Experimental Analysis of Waterproofing Polymeric Pavements for Concrete Bridge Decks

Marco Pasetto[1]/Giovanni Giacomello[2]

[1]Professor, University of Padua, Department of Civil, Environmental and Architectural Engineering - Padova, Italy

[2]Ph.D. student, University of Padua, Department of Civil, Environmental and Architectural Engineering - Padova, Italy

ABSTRACT

On concrete bridge and viaduct decks, traditional bituminous pavements are often subject to rapid degradation, particularly due to precipitation, traffic loadings and chemical attack. Pavement failure can also be due to underlying cracks related to steel reinforcement corrosion. For this reason waterproofing plays an important role in durability of the structure. Waterproofing can be done by means of polymeric binders and aggregates, mixed or applied together in the surface course.

The paper summarizes the main results of a study aimed at mechanically characterizing resin-aggregate mixtures (premixed and multi-layers) for bridge waterproofing and paving: two types of resins and several types of natural and artificial/industrial aggregate (EAF slag, C&D aggregate, limestone and quartz sand) were used. Permanent deformation resistance and adhesion tests were conducted, as well as trials to define the surface characteristics of the product (skid resistance, permeability, macro-texture). The comparison demonstrates that polymeric slurries present better resistance than polymeric multi-layers and bituminous mixtures to permanent deformation at higher temperatures (40 to 60 °C), but show some deficiencies in adhesion properties. However, the surface characteristics of polymeric multi-layers are preferable to those of slurries and traditional bituminous mixtures.

INTRODUCTION

Bridges and viaducts occur frequently along roads and motorways for crossing over watercourses, low-lying areas and other roads and railways. Many of these structures, due to the effect of increasing traffic loads and climatic stresses, have suffered from accelerated degradation over the years. This degradation has identified the need to study alternative maintenance techniques that are aimed at extending the service life of the surfacings (White and Montani, 1997).

Concrete and steel are the main materials used for constructing the deck of a reinforced concrete bridge. Both materials have characteristics of high resistance, which decrease in different ways. The porosity of the concrete allows the intrusion of de-icing salts and chemical agents that contribute to its deterioration and corrosion of the steel reinforcing. Moisture entering the concrete through the cracks and successive freeze-thaw cycles causes damage to the concrete and steel (Nabar and Mendis, 1997; Silfwerbrand and Paulasson, 1998).

The damages that these structures suffer over time are mainly due to precipitation, traffic loadings and chemical agents. The precipitation (rain and snow) and chemical agents can attack the steel and concrete and corrode them until parts of the structure fail. Water and snow, coming into contact with the steel, encourage the formation of rust, which increases the dimensions of the steel with the consequent opening of cracks in the concrete. The de-icing salts lead to the intrusion of chlorine ions, which accelerates the corrosion of the reinforcing (White and Montani, 1997).

The repetitiveness of the loads (by transiting vehicles) over time stresses the materials: this damage appears as cracks and fissures on the deck surface, which reduce the area of the section, leading to collapse.

The pavement on bridge decks therefore performs a twofold function: on the one hand it protects the concrete from chemical attack and weather, providing waterproofing; on the other it acts as a sort of filter/distributor of the vehicle loads. Sometimes the damage suffered by the concrete is reflected up through the pavement, which deteriorates over time with the transit of vehicles. Vice versa, the damage that the pavement suffers (permanent deformations, fatigue cracking, detachment of parts, etc.) involves damage to the deck. Cracks and detachments of the pavement allow the intrusion of water, snow and salt (during the winter), which damage the concrete and steel.

The pavements on bridge decks must, therefore, consist of materials with good mechanical properties in terms of surface characteristics, resistance to permanent deformation and fatigue, and adhesion to the support.

Many methods have already been utilized to improve these properties. Traditional systems exist composed of a bituminous layer that performs the function of waterproofing the deck. Two or three layers of asphalt are then laid on top of the waterproofing layer to reduce the tensile stress on the concrete induced by the traffic loads. According to some authors the laying of a waterproofing membrane between the interface between the concrete deck and the asphalt layer of the pavement is necessary in order to increase the water resistance and provide adequate adhesion at the interface between the layers (Zhou and Xu, 2009). Skid resistance can be improved by means of pigmented sand aggregate. Other authors suggest using asphalt mixtures with bitumen that has been modified by the addition of styrene-butadienestyrene (SBS), for trafficable waterproofings: rutting tests have indicated that traditional asphalt concretes suffer greater deformations than those modified with SBS (Park et al., 2009).

Some researchers argue that an asphalt pavement loses its surface characteristics (especially in terms of skid resistance) in about ten years and that detachments and loss of adhesion between pavement and deck occur. They, therefore, suggest using less costly and more long-lasting materials such as epoxy resins, cement concrete and concrete modified with latex (Babaei and Hawkins, 1988).

Other researchers suggest that a good bond of a pavement on concrete decks is to be ascribed to the roughness of the support. Utilizing suitable equipment, the deck should be roughened to allow a better grip of the coating layer (Silfwerbrand and Paulasson, 1998).

Yet others maintain that a good bonding of the coating, either a polymeric mixture or an asphalt, is to be ascribed to the presence of a thin layer that waterproofs the support (in epoxy resin or methacrylate) (Gillum et al., 2001). On pavements in asphalt or concrete, the laying of a thin layer of polymeric mortars to roughen the surface and obtain a good skid resistance between pavement and tyre has also been studied in Italy, with good results (Pasetto and Zanutto, 1999).

Another solution is the use of different types of polymers or synthetic resins (methacrylic, epoxy, polyurethane and polyester) as substitutes for the bituminous binder. Most of these binders have thermosetting behaviour: with a rise in temperature, the network of molecules that is created during the hardening is refined, forming other networks. Instead, bitumen tends to soften at high temperatures. These binders, therefore, perform better, especially in places with high temperatures, where the use of bitumen would lead to the exuding of the binder (bleeding).

Various studies have been dedicated to the laying methods of these polymeric mixtures, including in trials (Sprinkel, 1997; Knight et al., 2004; Pasetto and Giacomello, 2013). The first method is known as "premixed" and consists of mixing resin and aggregates in opportune quantities to form a mortar that is then laid on the deck (Calvo and Meyers, 1991; Maass, 2003). The second, called "multi-layer", is in a series of steps: 1) laying of a layer of binder (usually a few millimetres thick), 2) dusting of aggregate on the binder surface until saturation, 3) elimination of the aggregate in excess after setting, 4) laying of a new layer of binder (as in step 1) to then pass to steps 2 and 3. Just two layers of binder and aggregate are usually laid to form a pavement with pronounced surface characteristics. The final thickness normally reaches 8 mm to 10 mm (Sprinkel 2001; Stenko, 2001).

Some authors have demonstrated, with in situ trials, that the use of polymeric coatings based on epoxy resins confers many benefits: rapid intervention, low permeability, improvement of skid resistance, better ratio cost/useful life of the deck, excellent adhesion to the concrete support (not affected by the presence of alkalinity). The authors indicate that the system is suited to the repair and maintenance of old bridge decks. Even if the epoxy binder has a low modulus of elasticity, the pavement has good resistance and flexibility (Mendis, 1987; Dimmick, 1997; Nabar and Mendis, 1997; Zalatimo and Fowler, 1997).

The "premixed" method has been widely used in the USA to delay the process of corrosion in the concrete (preventing the penetration of chlorine ions and moisture): a waterproof coating of epoxy mortar has demonstrated, also 15 years after laying, a high resistance to wear and good maintenance of the surface characteristics, extending the service life of the pavement and deck (Dimmick, 1996).

The experimental programme presented in this paper studies the performances of materials intended for waterproofing surface courses applied to concrete bridge decks. The use of epoxy polymers has been evaluated, comparing the properties of different laying methods through tests of physical (surface characteristics) and mechanical characterization (permanent deformations and adhesion to the support).

MATERIALS

The materials used include various types of binders and marginal and natural aggregates (electric arc furnace steel slag, limestone, quartz sand, bauxite, etc.). The binders are two types of resin, two types of primer and one bitumen. Binder A (Table 1) is a two-component product (a monomer and a catalyst) of epoxypolyurethane, which hardens at ambient temperature. It can be laid alone to form a waterproofing layer (1 - 2 mm approx., following the "multi-layer" solution), or mixed with aggregates to form a mortar ("premixed" solution). Its principal characteristics are ductility, impermeability, tenacity and chemical inertia. The mortar possesses a flexural strength of 16 MPa and a compressive strength of more than 10 MPa (EN 196-1:2005). Binder B (Table 2) is a two-component (a monomer and a catalyst) modified polyurethane resin. This binder was used only for the multi-layer method.

Preliminary operations were involved prior to laying the binder or mortar, for both types (A and B). The surface of the support was first cleaned and sanded to eliminate any parts that were fragile or could break off. A layer of primer was then laid, which had the function of aiding the adhesion of the binder, or mortar, to the cement concrete support. Primer A (Table 1) is a synthetic epoxy material and possesses optimal resistance to water and saline solutions. Primer B (Table 2) is a twocomponent methacrylic resin, it is colourless and possesses low viscosity.

Specimens were also prepared with an asphalt mixture (BM), which was produced with a bitumen traditionally used in Italy (50/70 pen) (Table 3). To obtain a better adhesion to the concrete, a bituminous layer, composed of polyester covered in bitumen and chippings, with a waterproofing function, was interposed between the support and asphalt.

Tables 1 and 2 summarize the principal properties of the synthetic binders and primers. The physical-mechanical properties of the bituminous layer and bitumen are shown in Table 3.

Table 1: Properties of synthetic binder type A and primer type A

Parameter	Standard	Binder A	Parameter	Standard	Primer A
Density [Mg/m³]	ASTM D792	1,15	Slant shear strength [MPa]	EN 12615	> 12
Superficial hardness [Shore A]	ASTM D2240	60	Compressive strength [MPa]	ASTM D695	> 60
Tensile strength [MPa]	ASTM D638	≥ 2	Elastic Modulus in compression [MPa]	ASTM D695	> 2950
Tensile extensibility [%]	ASTM D638	≥ 100	Flexural-tensile strength [MPa]	ASTM D695	> 30

Strength of coating [MPa]	ASTM D4541	≥ 1,85	Bond strength by pull-off test [MPa]	EN 1542	> 5

Table 2: Properties of synthetic binder type B and primer type B

Parameter	Standard	Binder B	Parameter	Standard	Primer B
Density at 25°C [Mg/M³]	ISO 2811	0,99	Tensile strength [MPa]	ISO 527	13,8
Curing time at 20°C [s]	ISO 527	3600 – 7200	Elongation at max. strength [%]	ISO 527	1,3
Tensile strength at 20°C [MPa]	ISO 527	11	Modulus of elasticity [MPa]	ISO 527	1500
Elongation at Fracture, 20°C [%]	ISO 527	250	Elongation at fracture [%]	ISO 527	1,3
Elastic Modulus at 20°C [MPa]	ISO 527	82,4	Density at 20°C [Mg/m³]	ISO 1183	1,16

Table 3: Properties of bituminous layer and of bitumen (50/70 pen)

Parameter	Standard	BM	Parameter	Standard	Bitumen
Thickness [mm]	-	4	Penetration at 25°C [0.1 x mm]	1426	65
Extensibility [%]	ASTM D638	> 105	Softening point Ring & Ball [°C]	EN 1427	45
Tensile strength [MPa]	ASTM D638	> 2,5	Fraass breaking point [°C]	EN 12593	≤ -8
Bond strength by pull-off test [MPa]	UNI EN 1542	2, 5	Dynamic viscosity at 60°C [Pa · s]	EN 12596	≥ 145
Tensile modulus [MPa]	ASTM D638	> 9	Ductility at 25°C [mm]	ASTM D 113	≥ 800

The aggregates used are: quartz sand (QS), natural limestone (L), electric arc furnace steel slags (SS), recycled material (RM) and bauxite (B). The physical-mechanical properties of the aggregates are summarized in Table 4.

Table 4: Physical and mechanical characteristics of the aggregate: Quartz Sand (QS), EAF steel slag (SS), Limestone (L), Recycled material (RM), Bauxite (B)

Parameter	Standard	QS	SS		L		Rh/		B
Fractions	-		0/4	4/8	0 5	5/10	0/5	5/10	-
Los Angeles coefficient [%]	EN 1097-2	-	-	11,5	-	16	-	23,6	-
Equivalent in sand [%]	EN 933- 8	95	87	-	"0	-	98	-	98
Shape index [%]	EN 933- 4	-	19	15	-	5	15	22	-
Flakiness index [%]	EN 933- 3	-	29	8	-	8	16	32	-
Grain dry density [Mg/m³]	CNR 64/78	2,78	3,9	3,92	2,76	2,72	2,6	2,64	2,81
Grain bulk density [Mg/m³]	CNR 63/78	2,57	3,9	3,87	2,55	2,64	2,5	2,48	2,55
Aggregate bulk density [Mg/ m³]	CNR 62/78	1,48	2,1	2,13	1,48	1,36	1,3	1,32	1,52

The quartz sand and bauxite are very similar aggregates: they have a low quantity of fine particles and an intermediate specific weight between the steel slags and recycled material. Two grading fractions were used for the natural limestone and recycled aggregate: 0/5 and 5/10 mm. Two were also used for the steel slags: 0/4 and 4/8 mm.

The steel slags have the highest specific weight, a low content of fine particles and high abrasion and Los Angeles fragmentation resistance (low value of the Los Angeles coefficient). A low value of the flakiness and shape index, as in the case of the limestone and steel slags, indicates a low presence of aggregates of lenticular and elongated shape: greatly sought after qualities to obtain a good asphalt for a surface course. The recycled material, a mixture of glass and material from construction and demolition, has properties with among the highest values: low resistance to fragmentation and abrasion and a more elongated shape of the aggregates than the other materials.

The studied solutions regard both the "premixed" (PR) and the "multi-layer/multiplelayer" (ML) laying methods. The mixtures prepared according to the "premixed" method were further divided into two families, depending on the grading envelope utilized for the mix design: the type 1 (T1) envelope, already used (Pasetto and Giacomello, 2013) for the laying of mixtures, and the type 2 (T2) envelope used for an asphalt for surface courses (SITEB, Italian Society of Bitumen Technicians).

With the first type of envelope (T1) polymeric mixtures were prepared with quartz sand (PR-T1-QS), steel slags (PR-T1-SS) and recycled aggregate (PR-T1-RM). With the second type of envelope (T2) polymeric conglomerates were produced with limestone (PR-T2-L) and steel slags (PR-T2-SS). Only resin of type A was used for the preparation of both types. For the preparation of the asphalt, two systems were adopted that are traditionally used in northern Italy: one with limestone (BM-T2-L) and one with steel slags (BM-T2-SS). For the multiple-layer type, a grading envelope from the literature (Pasetto and Giacomello, 2013) was used and polymeric mixtures were prepared with quartz sand (ML-QS), steel slags (ML-SS), recycled aggregate (ML-RM) and bauxite (ML-B). The type B binder was only used to produce ML-B specimens. The proportions between the grading fractions used are indicated in table 5. The grading of the mixtures and grading envelopes are reported in figures 1, 2 and 3.

Table 5: Aggregate type and particle size distribution of the mixtures - composition of the mixtures

Mixture composition	Fraction [mm]	Quantity [%								
		BM-T2-L & PR-T2	BM-T2-SS & PR-T2-SS	PR-T1-RM	PR-T1-QS	PR-T1-SS	ML-QS	ML-SS	ML-RM	ML-B
Limestone	0/5	48	-	40	-	-	-	-	-	-
	5/W	45	-	-	-	-	-	-	-	-
EAF steel slags	0/4	-	70	-	-	100	-	100	-	-
	4/8	-	22	-	-	-	-	-	-	-
Recycling material	0/5	-	-	45	-	-	-	-	100	-
	5/10	-	-	15	-	-	-	-	-	-
Quartz sand 1		-	-	-	100	-	-	-	-	-
Quartz sand 2		-	-	-	-	-	100	-	-	-
Bauxite		-	-	-	-	-	-	-	-	100
Mineral filler (additive)		7	8	-	-	-	-	-	-	-

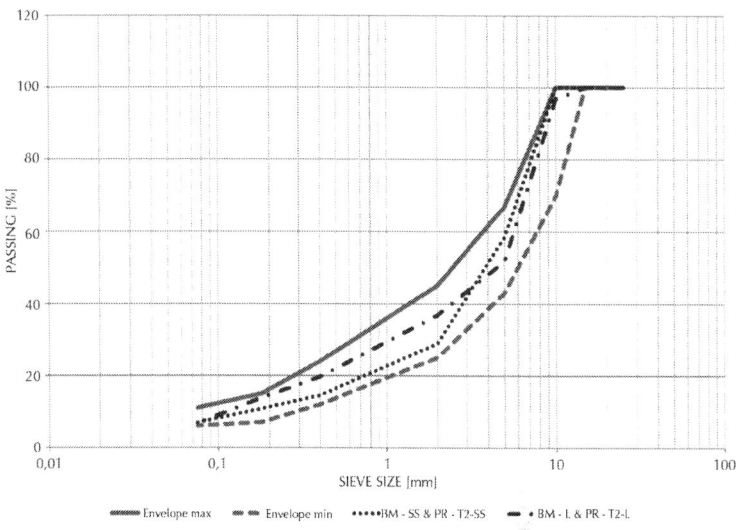

Figure 1: Grading envelope and curves for BM-T2-SS,BM-T2-L, PR-T2-L e PRT2- SS mixtures (asphalt mixtures BM and premixed waterproofings PR)

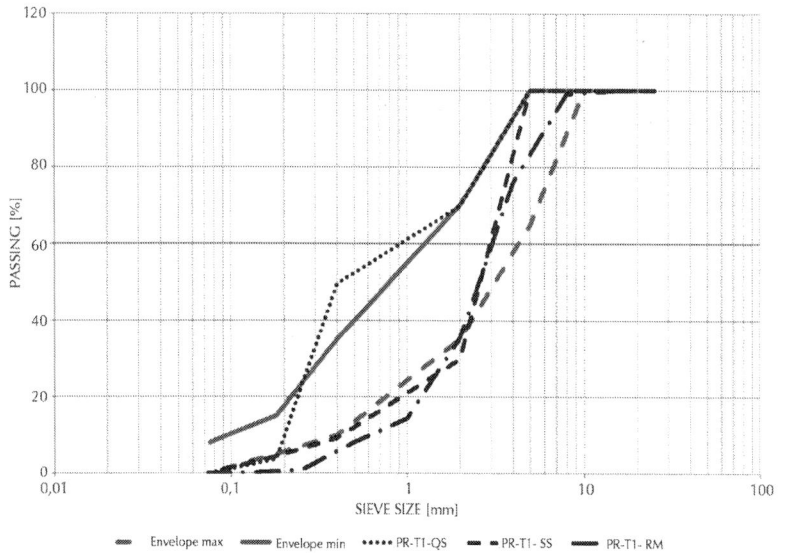

Figure 2: Grading envelope and curves for PR-T1-QS, PR-T1-SS e PR-T1-RM mixtures (premixed waterproofings PR)

The mixtures with the "premixed" laying type were produced following a precise procedure. After having cleared impurities from the surface of the concrete support, the primer was prepared (mixing a monomer and a catalyst) and the mortar (initially mixing the binder with an adequate amount of catalyst and then also with the aggregate). Having laid the primer, the polymeric coating was added after an interval of a few minutes. The specimen was then left to rest for a day to allow the mortar to set.

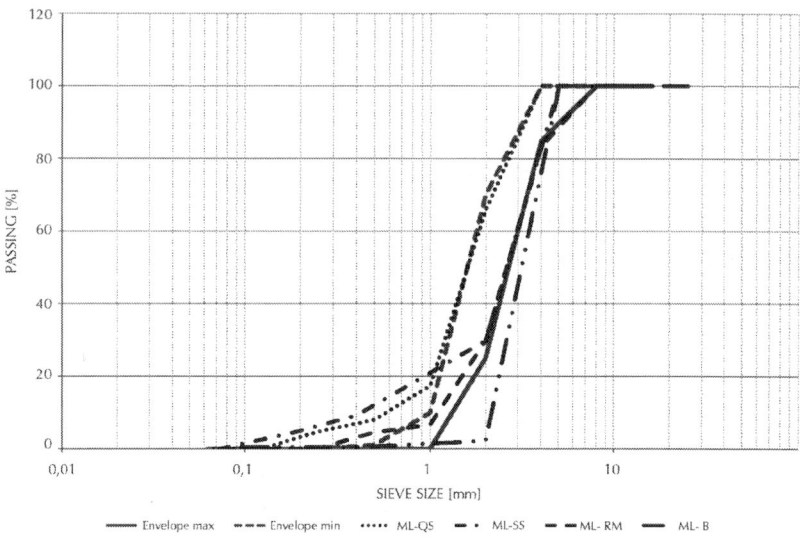

Figure 3: Grading envelope and curves for ML-QS, ML-SS, ML-RM e ML-B mixtures (multi-layer waterproofings ML)

In the case of mixtures with the "multi-layer" type of laying the procedure followed was slightly different. Having cleaned the surface of the support, prepared and laid the primer, a thin layer of binder (obtained mixing resin and catalyst) was laid. Before the binder set, the aggregate was dusted onto the surface until it was saturated. After setting (one day of rest), the excess aggregate was removed with a blast of air. A second layer of binder was laid on the aggregate remaining stuck to the support, and the aggregate re-applied, until the surface was saturated. After another day of rest, the polymeric pavement was ready.

Just a single layer of aggregate was scattered in the case of the type B resin, because there is not only the primer beneath the binder, but also another layer of waterproofing resin. The concrete supports are 300 mm long, 400 mm wide and 30 mm thick. However, the final thickness of the specimen varies according to the type of coating: in the "multi-layer" case a pavement can be 15 mm, while in the "premixed" case it may reach approx. 20 mm (like the pavement in asphalt).

METHODS

The first tests conducted on the specimens were the measurements of surface characteristics. These are very important for pavements on bridges with a concrete deck because they must offer optimal adhesion between tyre and road surfacing. The macro-texture, by means of the height in sand (EN 13036-1:2010), skid resistance (PTV - Pendulum Test Value index) by the PendulumTest (EN 13036-4:2011), and permeability (EN 13036-3:2006) were, therefore, evaluated.

The macro-texture of the pavement was measured according to the protocol indicated in the EN standard. Known quantities of microspheres are spilled onto the pavement and spread to form a circular area, until all the elements have filled the voids between aggregates. The two diameters of the circular area are then measured and, the volume of spheres being known, the macro-texture of the pavement is calculated in terms of height in spheres.

The skid resistance is measured by a rubber sliding block of determined characteristics connected to a pendulum that oscillates without friction nominally around a pivot. The sliding block, made to fall from a precise height, rubs against the surface of the pavement and loses a certain amount of kinetic energy. The measurement taken with this instrument expresses the amount of energy lost after the rubbing of the sliding block on the pavement: the smoother the pavement is the lower the value of skid resistance will be.

The permeability was measured as the time (in seconds) in which a known quantity of water, contained in a test-tube, leaks from beneath a rubber sealing ring pressed by a mass (of known weight).

The specimens also underwent rutting (wheel tracking) tests (EN 12697-22:2007) using the small device to Procedure B. The test simulates the repeated transit of heavy vehicles on the pavement, to register the permanent deformation over time. The number of transits (104 cycles) simulated that of the average traffic on Italian roads by means of a rubber wheel (assimilating a vehicle tyre). The tests were done at temperatures of 0 °C, 20 °C, 40 °C and 60 °C, in order to understand the thermosetting behaviour of the epoxy resins.

In all cases the adhesion of the pavement to the concrete support was tested. The standard EN 1542:2000 required cylindrical specimens to be produced of approximately 50 mm in diameter milling the material with a coring rig, to a depth of approximately 15mm into the underlying concrete layer. A round peg was then glued onto each specimen with epoxy adhesive (prescribed by the standard), on which tensile stress was exerted until the specimen failed. The protocol prescribes that the tension increase must be progressive and no more than 0.05 MPa/s.

RESULTS AND DISCUSSION

The tests to establish the surface properties of the pavements were conducted on each coating, doing three measurements with the macro-texture by patch method, five with the Pendulum Test and three for the permeability. The results are shown in Table 6.The tests for measuring the surface characteristics provided clear indications, even during the phase of producing the pavements, especially in the cases of the multi-layer specimens. In any case the surface characteristics vary according to the type of aggregate used and the laying method.

The coatings produced with the multi-layer method, compared with those of the premixed method, provide much more satisfactory results in terms of surface characteristics: the macro-texture and skid resistance are much better. Almost all the pavement types show a macro-texture of the "average" type for Italy (values between 0.40 and 0.80 mm in terms of height) and only the multi-layer coatings arrive at macro-textures of the "very rough" type (values of above 1.20 mm). These values make sense observing that the aggregates create much greater roughness when the multi-layer method is used (Figure 4).

Table 6: Mean value of height in sand, Pendulum Test skid resistance and permeability

Coating type	PR-T1-QS	PR-T1-SS	PR-T1-RM	PR-T2-SS	PR-T2-L	BM-T2-L	BM-T2-SS	ML-B	ML-QS	ML-SS	MLRM
Height in sand [mm]	0,32	0,60	0,58	0,28	0,36	0,72	0,52	0,46	1,54	2,11	2,31
Mean PTV [-]	21	71	86	31	36	116	95	76	104	114	99
Permeability [s]	>180	>180	585	>180	>180	163	184	742	21	18	14

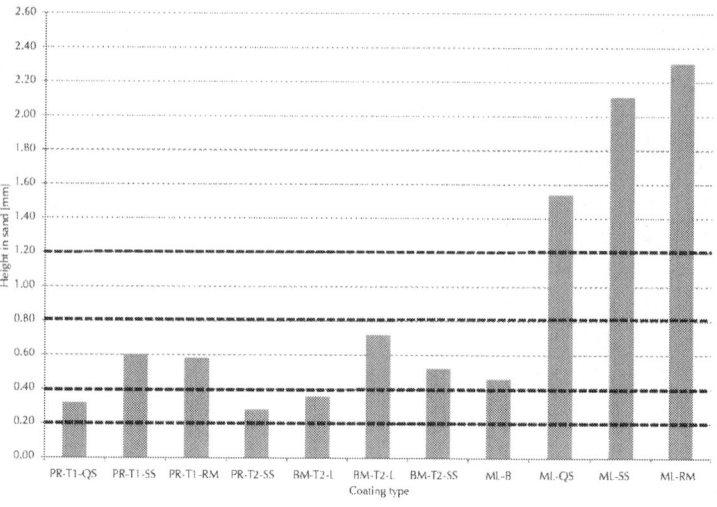

Figure 4: Mean value of macro-texture for each coating type

In terms of skid resistance, almost all types of coating have a surface that is "antiskid par excellence" (values above 65). In the same way as the macro-texture, the multilayer coatings and asphalt have better skid resistance (Figure 5).

The recycled material has good roughness in both cases (premixed and multi-layer waterproofing) and is an antiskid surfacing par excellence (in terms of skid resistance as defined in Italy). The steel slags instead lead to passable results. The coatings with a type 2 lithic matrix have good surface characteristics if bitumen is used as the binder, but not with the polymeric binder. This is confirmed visually: the resin tends to fill the voids between the aggregates creating a rather smooth surface, whereas the bitumen, covering the single grains of aggregate, allows them to retain their shape. In this way, the global properties of the bituminous pavement are better than those with a polymeric binder.

The permeability measured confirms what has already been described. The coatings with smoother surfaces impede water seepage and therefore have lower values of permeability. Conversely, the multi-layer pavements, which maintain greater roughness of the aggregates, allow better water drainage. The multi-layer with bauxite aggregate coating has values intermediate between those of the asphalt and those found on pavements of the premixed type (table 6).

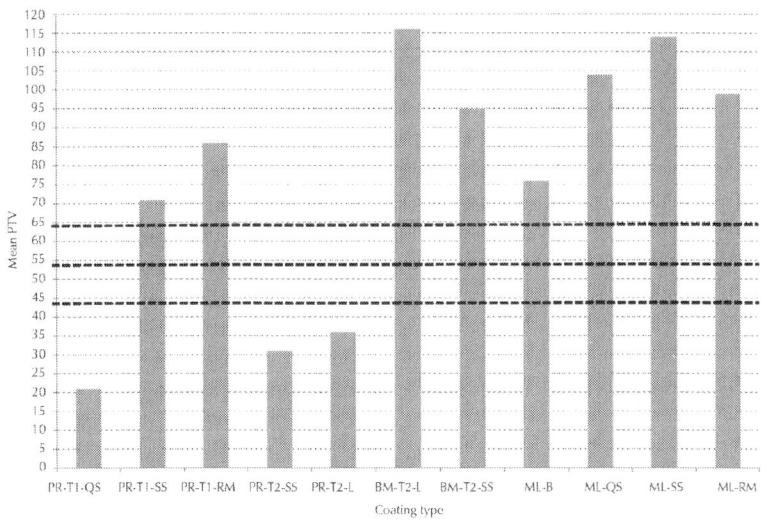

Figure 5: Mean value of skid resistance for any coating type

The results of the tests with the wheel tracking device are reported in Figures 6 and 7.

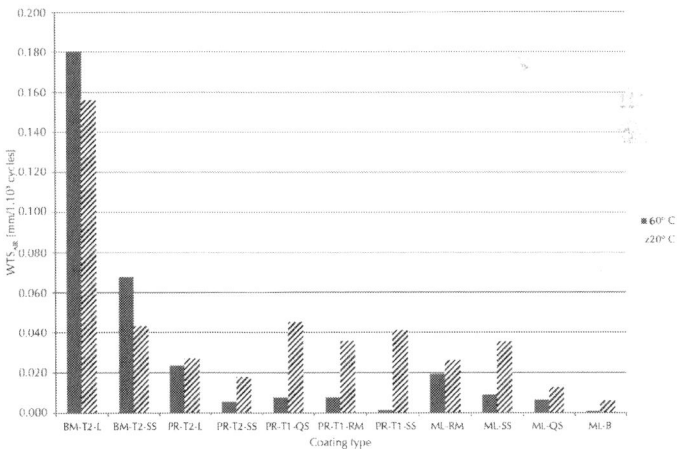

Figure 6: WTSAIR at 20°C and 60°C for each coating type

In each test, the parameter WTS_{AIR} was obtained (as reported in the EN standard), which is the slope of the curve of deformation in millimetres per 103 loading cycles. The calculation consists in subtracting the value of the height of the deformation at 5,000 cycles from that at 10,000 cycles, and dividing the result by 5.

It was possible to ascertain immediately that the polymeric binders possess thermosetting behaviour, due to their chemical composition. In fact, the network of molecules becomes complete, stiffening the polymeric matrix, with the increasing of the temperature.

The polymeric coatings with type 2 lithic matrix react better to permanent deformations than the bituminous ones. In addition, the asphalt with slags performs better than that with limestone, as found previously (Pasetto and Baldo, 2012). The multi-layer polymeric coatings suffer more from an accumulation of permanent deformations than the same coatings of the premixed type (Figure 6).

In all the cases analysed, the polymeric coatings (premixed and multi-layer) have shown a low predisposition to permanent deformation at high temperatures (60 °C – 40 °C) and, vice versa, a high predisposition at lower temperatures (0 °C – 20 °C) (Figure 7).

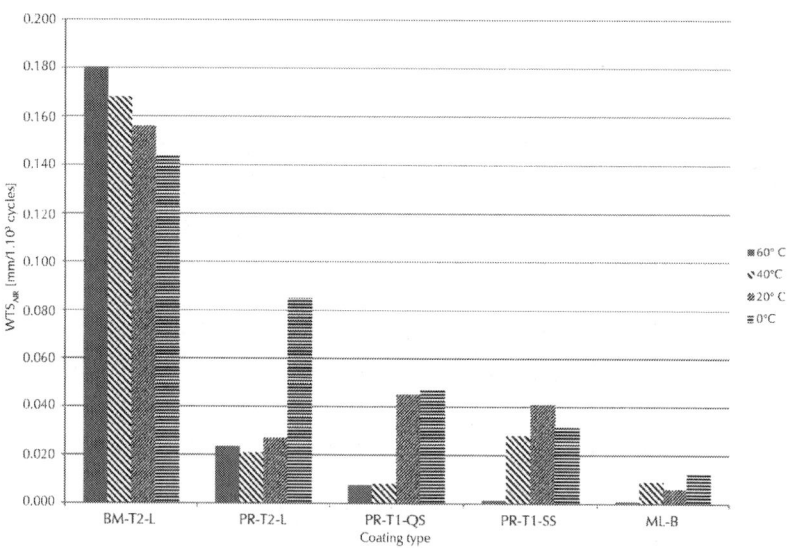

Figure 7: WTS_{AIR} at 0°C, 20°C, 40°C and 60°C for several coating types

Lastly, the adhesion of the coating, polymeric and non-polymeric, to the concrete support was tested. The averages of three tests of adhesion were done for each type of pavement. The results, summarised in table 7, differed each time, even with the same specimen. This occurred mainly because the surface of the coating presented discontinuities that impeded the perfect attachment of the adhesive. In addition, even a slight inclination of the peg or the instrument led to non-homogeneous forces of traction on the specimen.

Only in the case of the premixed type 2 envelope with steel slag waterproofing, did the failure of the specimen occur inside the concrete support, evidencing that the type 2 matrix with slags has a good internal cohesion. In fact, even if the stresses developed are high, the primer or the interface between primer and coating has almost always given way.

The asphalts have given a poorer performance than the other coatings: indeed, the bituminous layer was shown to have less adhesion to the concrete support than the primers used for the other solutions. In two cases, during the core sampling, just the rotation of the coring rig led to the waterproofing layer and the asphalt detaching from the support.

Table 7: Value of tensile stress and type of failure for each coating type

Coating type	Mean tensile stress [MPa]	Type of failure
PR-T2-L	1,44	within the adhesive layer of the dowel
PR-T2-SS	1,39	within the concrete substrate
BM-T2-L	0,70	at the interface between the concrete substrate and the bituminous membrane
BM-T2-SS	0,76	within the bituminous mixture
PR-T1-QS	1,59	at the interface between the concrete substrate and the primer
PR-T1-SS	1,31	within the primer
PR-T1-RM	0,91	within the primer
ML-QS	1,53	at the interface between the primer and the coating layer
ML-SS	1,48	within the primer
ML-RM	0,97	within the coating layer
ML-B	1,20	the interface between the concrete substrate and the primer

The values of strength are globally higher for the solutions laid with the multi-layer method than with the premixed method, even if the solutions with quartz sand both obtained high values of tensile strength.

CONCLUSIONS

The research aim was to study pavements made with polymers for concrete decks on bridges. The performance of these coatings has been evaluated, through laboratory tests and data analysis, from the physical (surface characteristics) and mechanical (permanent deformations and adhesion to the support) points of view.

Overall, in terms of both macro-texture and Pendulum Test value, the multi-layer solution provides better surface characteristics than both the premixed solution and the traditional ones in asphalt.

This comparison suggests that coatings can be produced with the multi-layer system with better surface characteristics, but with a higher accumulation of permanent deformations than the premixed method. The cause of this behaviour is due to the aggregate grains not being completely immersed in the resin. In fact, the very rough surface of the coating confers antiskid properties on the system even in the worst cases, but renders the overall pavement weaker because of a greater presence of voids and the stresses that the tyre treads discharge directly on the single grains of the aggregate.

The premixed mixtures, therefore, have the strong point of being more resistant to high temperatures (that derives from the behaviour of the thermosetting resin used as binder), unlike the asphalts, which accumulate more permanent deformations. The performances obtained with quartz sand (QS) and steel slags (SS) are better than those with recycled material (RM). In fact, this last type of aggregate develops less adhesion to the epoxy resin because it has a higher flakiness coefficient than the other types of aggregate.

In addition, it has been found that the use of a type 2 lithic matrix, similar to that used for an asphalt, with polymeric binders leads to worse characteristics than the premixed solutions. For these last types, a type 1 lithic matrix is used with a grading envelope that includes only fine material. The union of polymers and fine aggregate in the premixed solution has instead conferred poorer surface characteristics, but a lower accumulation of permanent deformations. The results of the adhesion tests suggest the need for further studies.

REFERENCES

1. BABAEI, K. & HAWKINS, N.M. (1988) "Evaluation of bridge deck proctetive strategies", *ACI Concrete International*, Vol. 10, No. 12, pp 56-66.

2. CALVO, L. & MEYERS, M. (1991) "Overlay materials for bridge decks", *ACI Concrete International*, Vol. 13, No. 7, pp 48-49.

3. CARTER, P.D. (1997) "A procedure for determining performance of thin polymer overlays on Alberta bridge decks", *ACI Special Pubblication*, Vol. 169, pp 107-121.

4. DEPUY, G.W. & DIMMICK, F.E. (2003) "Polymer concrete overlay for the repair and protection of concrete", *ACI Special Pubblication*, Vol. 214, pp 139-150.

5. DIMMICK, F.E. (1996) "15-year tracking study: comparing epoxy polymer concrete to Portland cement concrete applied on slab-on-grade and bridge decks", *ACI Special publication*, Vol. 166, pp 211-231.

6. DIMMICK, F.E. (1997) "Premixed epoxy polymer concrete bridge deck overlays", *ACI Special Pubblication*, Vol. 169, pp 146-171.

7. GILLUM, A.J., SHAHROOZ, B.M. and COLE, J.R. (2001) "Bond strength between sealed bridge decks and concrete overlays", *ACI Structural Journal*, Vol. 98, No. 6, pp 872-879.

8. KNIGHT, M.L. et al. (2004) "Overlay types used as preventive maintenance on Tennessee bridge deck", *Transportation Research Record: Journal of the Transportation Research Board*, National Research Council, Washington, D.C., 1866, pp 79-84.

9. MAASS, J. (2003) "How polyester polymer concrete highway and bridge deck overlays became state of the art", *ACI Special Pubblication*, Vol. 214, pp 39-50.

10. MENDIS, P. (1987) "A polymer concrete overlay", *ACI Concrete Internatinal*, Vol. 9, No. 12, pp 54-56.

11. NABAR, S. & MENDIS, P. (1997) "Experience with epoxy polymer concrete bridge deck thin overlays in service for over 10 year", *ACI Special Pubblication*, Vol. 169, pp 1-17.

12. PARK, H.M. et al. (2009) "Performance evaluation of a high durability asphalt binder and a high durability asphalt mixture for bridge deck pavements". *Construction and Building Materials*, Elsevier Ltd., Vol. 23, pp 219-225.

13. PASETTO, M. & ZANUTTO, G. (1999) "Irruvidimenti superficiali con inerti e resine", *Le Strade*, La Fiaccola, Milano, 4, pp 72-76.

14. PASETTO, M., MODENA, C., SILVAN, S. and BRUNO P. (2000) "Impermeabilizzazione e pavimentazione di ponti e viadotti con trattamentiirruvidenti a base di malta sintetica", Proceedings of X National S.I.I.V. Congress, 26- 28 October, Catania, Italy, C.16 paper, p. 11.

15. PASETTO, M. & BALDO, N. (2012) "Performance comparative analysis of stone mastic asphaltswith electric arc furnace steel slag: a laboratory evaluation". *Materials and Structures*,Vol. 45, pp 411–424.

16. PASETTO, M. & GIACOMELLO, G. (2013) "The use of synthetic resins on concrete bridge decks", Proceedings of 17th International Conference on Environemental and Mineral Processing, 6-8. 6. 2013. VSB - Technical University of Ostrava, Czech Republic.

17. SILFWERBRAND, J. & PAULASSON, J. (1998) "Better bonding of bridge deck overlays", *ACI Concrete International*, Vol. 20, No. 10, pp 56-61.

18. SPRINKEL, M. (1997) "Nineteen year performance of polymer concrete bridge", *ACI Special Pubblication*, Vol. 169, pp 42-74.

19. SPRINKEL, M. (2001) "Maintenance of concrete bridges", *Transportation Research Record: Journal of the Transportation Research Board*, National Research Council, Washington, D.C., Vol. 1749, pp 60-63.

20. STENKO, M.S. & CHAWALWALA, A.J. (2001) "Thin polysulfide epoxy bridge deck overlays". *Transportation Research Record: Journal of the Transportation Research Board*, National Research Council, Washington, D.C., Vol. 1749, pp 64-67.

21. ZALATIMO, J.-A.H. & FOWLER, D.W. (1997) "5-year performance of overlays in Fort Worth, Texas", *ACI Special Pubblication*, Vol. 169, pp 122-145.

22. ZHOU, Q. & XU, Q. 2009 "Experimental study of waterproof membranes on concrete deck: interface adhesion under influences of critical factors". *Materials and design*, Elsevier Ltd., Vol. 30, pp 1161-1168.

23. WHITE, D. & MONTANI, R. (1997) "Thin-bonded polymer concrete overlays for exposed concrete bridge deck protection and maintenance", *ACI Special Pubblication*, Vol. 169, pp 99-106.

Thermal Stability Analysis under Embankment with Asphalt Pavement and Cement Pavement in Permafrost Regions

Zhang Junwei[,1,2] Li Jinping,[3] and Quan Xiaojuan[4]

[1]School of Civil Engineering and Architecture, Southwest Petroleum University, Chengdu 610500, China

[2]Key Laboratory of Structure Engineering of Universities in Sichuan, Chengdu 610500, China

[3]CCCC First Highway Consultants Co., Ltd, Xi'an 710075, China

[4]School of Civil Engineering, Southwest Jiaotong University, Chengdu 60031, China

ABSTRACT

The permafrost degradation is the fundamental cause generating embankment diseases and pavement diseases in permafrost region while the permafrost degradation is related with temperature. Based on the field monitoring results of ground temperature along G214 Highway in high temperature permafrost regions, both the ground temperatures in superficial layer and the annual average temperatures under the embankment were discussed, respectively, for concrete pavements and asphalt pavements. The maximum depth of temperature field under the embankment for concrete pavements and asphalt pavements was also studied by using the finite element method. The results of numerical analysis indicate that there were remarkable seasonal differences of the ground temperatures in superficial layer between asphalt pavement and concrete pavement. The maximum influencing depth of temperature field under the permafrost embankment for every pavement was under the depth of 8 m. The thawed cores under both embankments have close relation with the maximum thawed depth, the embankment height, and the service time. The effective measurements will be proposed to keep the thermal stabilities of highway embankment by the results.

INTRODUCTION

Permafrost, a seasonal active layer sensitive to the temperatures, has been attracting deeply more and more researchers in the worldwide because of the permafrost degradation directly threating to the stability of building foundations. In particular, for the high temperature permafrost regions, all kinds of pavement diseases of the embankment would appear easily if the constructed embankments were not properly protected before. For example, there is the unevenly transverse settlement of pavement due to the asymmetric distribution of the thermal field under the permafrost embankment. For this reason, the thermal stability of cement concrete pavement in permafrost regions has been always studied [1–4]. The thermal state difference between south faced slope and north faced slope of embankment is considered as the most important factor to cause the unevenly transverse settlement of pavement. As a result, the thermal state difference may result in asymmetric thermal regime in embankment. And the unevenly transverse deformation may appear thereupon. In opinions of Yu [5], the permafrost degradation is mainly caused by the natural thermal equilibrium destroyed under permafrost embankment [6–11]. Affected by global warming and embankment construction disturbances, the heat exchange between atmosphere and ground surface had been changed after highway pavements were built in the permafrost regions. The permafrost degradation acceleration begins to grow due to the heat exchange condition changing. In particular, for warm permafrost regions, all kinds of pavement diseases would appear easily because of the acceleration of permafrost degradation if the constructed embankments were not properly protected. The previous studies on the acceleration of permafrost degradation indicate that embankment diseases and pavement diseases are closely related to thermal stability under permafrost embankment in the permafrost regions. It is a key factor to keep the embankment thermal stability to solve the problems of embankment diseases and pavement diseases. The unevenly distribution of the thermal field under the permafrost embankment is the direct reason of the embankment deformation in permafrost regions of the Qinghai-Tibetan Plateau. The maximum deformation of highway permafrost roadbed lies off the center of the embankment [5, 12].

To thoroughly understand the failure mechanisms of the Qinghai-Tibetan highway embankment in the permafrost regions, radiation and thermal balance were observed on the surface of asphalt pavement. And these observation stations were located on ground surfaces of different types of permafrost embankments between the Kunlun Pass and the No.66 station port. The calculated results based on the observation suppose that the part of heat influencing significantly on the heat regime of the embankment is considered as the main reason of the Qinghai-Tibetan highway embankment failure. And the part of heat was responsible for the formation of a thawed core below the embankment in the high-temperature permafrost section [13]. From the geohazard investigations of the embankments along the Qinghai-Tibet Railway, permafrost degradation results in some main geohazards such as thawing settlement, frost-heave, and freezing-thawing induced hazards. All of them might potentially influence the embankment stability including settlement, burying, and laterally thermal erosion [3]. The thermal stability of the permafrost embankment will highly have been changed because thermal effect problems associated with slope orientations result in the maximum thawed depth position being deviated in the roadbed transverse direction rather than thawed on the embankment central line [4]. The sunny-shady slope seriously impacts on the thermal and deformation stability of the highway embankment in warm permafrost region by analyzing both the observational geotemperature and deformation data of the embankment in the experimental section K369+100 along the National Highway 214 on Qinghai-Tibet Plateau [14]. From the insitu geothermal data of up to 15 years of 5 typical sections along the Qinghai-Tibet Highway, embankment settlement is closely related to the process of degrading of underling permafrost extensively in the five typical sections with different characteristics. With the increase in the mean annual ground temperature, the thawing rate firstly increases and then fluctuates as temperature rise rate increases and then decreases at the mean annual ground temperature of about −0.5°C [15]. Different from the previous study, in cold seasons, the temperature under the reinforced concrete component was higher than the shoulders by analyzing temperature characteristics of tested embankments at the Changchun site of Harbin to Dalian Passenger Dedicated Line. This difference decreases with the depth of roadbed. In warm seasons, these phenomena appear as a reverse trend, and also the temperature difference decreases with the depth of roadbed as usual. In different parts of the roadbed, the maximum seasonal frozen depths were all higher than the natural ground. The reason is that the roadbed materials changed the heat exchange process between the air and the ground surface [16]. Thus thaw settlement is the main embankments distresses of highway in permafrost regions, according to survey data of the Qinghai-Tibet Highway. It can be effectively mitigated or even controlled by raising the embankment height [17].

It is found from the aforementioned studies that cement concrete pavement and asphalt concrete pavement are presently two common pavement structures used in permafrost embankment engineering. Black asphalt pavement has strong temperature susceptibility. And it can absorb more solar radiation than cement concrete pavement. So the black asphalt pavement has a higher road surface temperature. This will seriously affect the temperature field and the stability of permafrost under the asphalt concrete pavement embankment. The aforementioned studies only consider the temperature field under asphalt pavement permafrost embankment with strong temperature susceptibility. Although some measures were used to deal with the embankment diseases and the pavement diseases, the disease problems could not be solved fundamentally. The most important reason is that the thermal stability under the permafrost embankment has not been quite figured out. As a result, cement concrete pavement begins to be selected to use to keep the temperature field of embankment stable instead of black asphalt pavement in the permafrost regions. Because the cement concrete pavement is seldom used under the natural conditions in the permafrost regions until now, it is not almost seen on the theory of cement concrete pavement or embankment diseases. In addition, these diseases are closely related with not only pavement structures but also the temperature field under both permafrost embankments. Therefore, the finite element method will be applied to calculate temperature field under both permafrost embankments with adjustable embankment height. The temperature field changes between asphalt pavement and cement concrete pavement will be analyzed at different embankment width. The studying results will supply reliable theory for the designing of the embankment stability. They are in favor of sustainable development of permafrost engineering because of keeping the thermal stability under the embankment by taking effective use of natural ventilation.

ANALYSIS OF THE SHALLOW GROUND TEMPERATURE

The cement section of K374+975 and the asphalt section of K375+300 are the borders upon sections. The ground temperature of the both pavements has being observed since July 2003 Figure 1 shows that ground temperature at the depth of 0.5 m versus. time under the three pavements types from August 1, 2003 to August 1, 2006 according to the observation results of the ground temperature. These observation results were mainly obtained through the various drill-holes by the geological drilling. And we can find that there is an obvious ground temperature difference between the section of K374+975 and the section of K375+300. As is shown in Figure 1, the construction of both the cement pavement and the asphalt pavement caused a large change in the shallow thermal regime partly. There is also distinct difference in the ground temperature at the depth of 0.5 m between both pavements. The ground temperature under cement pavement is always lower than that under asphalt pavement. And there is distinctly seasonal difference in the ground temperature at the depth of 0.5 m between them. It is obviously larger in summer than in winter. In summer, the ground temperature under the cement pavement is higher 5-6°C than that under the natural ground at the depth of 0.5 m. However, the ground temperature under the asphalt pavement is even higher 10°C than that under the natural ground at the depth of 0.5 m. In winter, the difference between them is little. Both of them are less than 1.0°C. The reason is that the temperature sensitivity of the asphalt is higher than that of the cement.

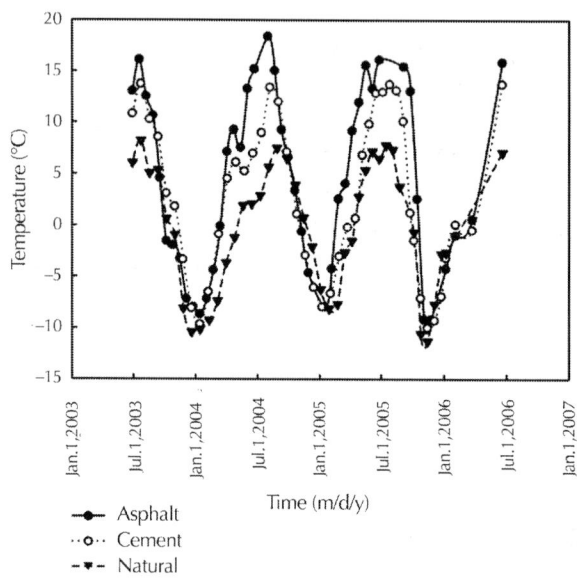

Figure 1: The ground temperature at the depth of 0.5 m versus time.

In order to further narrate the ground temperature difference between both pavements, Table 1 gives the annual average temperature of 2003–2006 years under asphalt pavement in the section K417+970 of seasonally frozen ground and cement pavement in the section K418+030 of permafrost, as well as the difference between them at the depths of 0.5–2.0 m. It is seen from Table 1 that the annual average temperature under asphalt pavement in the section K417+970 of seasonally frozen ground at the depths of 0.5–2.0 m is higher (1.84–2.53°C) than that under cement pavement in the section K418+030 of permafrost from 2004 to 2006. The ground temperature difference between both pavements in the section of seasonally frozen ground is even larger than that in the section of permafrost. It is also shown that constructed cement pavement is more stable than constructed asphalt pavement in the thermal stability of embankment.

Table 1: The ground temperatures difference under asphalt pavement and cement pavement (the depth of 0.5 m)

Time	Ground temperature difference between asphalt and natural pavements/°C	Ground temperature difference between asphalt and cement pavements/°C
2004	3.93	1.84
2005	4.43	2.36
2006	3.71	2.03
The average temperature of 03–06 years	4.02	2.06
The average temperature Zhu Linnan supplied	4.0	2.0

As can be seen from Table 2, the ground temperature at the depth of 2.0 m under the asphalt pavement is 0.47–0.92°C higher that than under cement pavement. This result manifests good thermal stability of permafrost under cement pavement. There is also a temperature difference between the asphalt pavement and the cement pavement at the same depth under the pavement. And the difference reduces gradually with the increment of the depth.

Table 2: Comparisons of the ground temperature at the depth of 2.0 under pavement

Depth/m	Asphalt pavement	Cement pavement	Ground temperature difference/°C
	The average temperature of 03–06 years/°C	The average temperature of 03–06 years/°C	
0.5	2.69	1.77	0.92
1	2.5	1.6	0.9
1.5	2.05	1.43	0.62
2	1.61	1.14	0.47

It can be found from Table 3 that the ground temperature of the embankment centre at the depth of 2.0 m under the asphalt pavement is 0.68–1.34°C higher than that under cement pavement. The ground temperature of the embankment centre at the depth of 2.0 m–8.0 m under the asphalt pavement is 0.06–0.29°C higher than that of cement pavement. But the temperature difference began to become less and less beyond 8 m under the ground surface. This result manifests that for the embankment with a width of 8.5 m, 8 m is the maximum depth of the pavement inflecting on embankment. Otherwise, the ground temperature under the embankment is influenced little on beyond 8 m.

Table 3: The average temperatures of 03–06 years under asphalt pavement and cement pavement

Depth/m	Cement pavement/°C			Cement pavement/°C			Ground temperature difference/°C
	Center hole	Natural hole	Temperature increment magnitude	Center hole	Natural hole	Temperature increment magnitude	
−1.3	2	/	2.0	3.26	/	3.26	1.26
−0.8	2.11	/	2.11	3.45	/	3.45	1.34
−0.3	1.82	/	1.82	2.89	/	2.89	1.07
0.2	1.59	/	1.59	2.27	/	2.27	0.68
0.7	1.15	−0.53	1.68	1.65	−0.32	1.97	0.29
1.2	0.72	−0.35	1.07	1.02	−0.32	1.34	0.27
2.2	0.10	−0.47	0.57	0.14	−0.67	0.81	0.24
3.2	−0.20	−0.57	0.37	−0.19	−0.68	0.49	0.12
4.2	−0.30	−0.58	0.28	−0.35	−0.71	0.36	0.08
5.2	−0.36	−0.59	0.23	−0.45	−0.75	0.3	0.07
6.2	−0.40	−0.59	0.19	−0.53	−0.78	0.25	0.06
7.2	−0.42	−0.60	0.18	−0.59	−0.80	0.21	0.03
8.7	−0.46	−0.59	0.13	−0.67	−0.82	0.15	0.02

Both the most high ground temperate and the lowest ground temperature at the depth of 2.0 m under the pavement in Table 4 are from the section K374+975 and the section K375+300 in the national highway 214 for the natural ground, cement pavement, and asphalt pavement, respectively. But the curve of the surface temperature amplitude under different pavements in Figure 2 is not based on Table 4.

Table 4: The temperature of linear regression in surface of natural ground, asphalt pavement, and cement pavement

Monitoring section	Pavement type	Monitoring contents	04	05	06
K374+975	Natural ground	The maximum temperature	11.22	10.32	10.87
		The minimum temperature	−12.85	−12.77	−13.31
		The annual average ground temperature	12.03	11.54	12.09
		The average temperature amplitude of 04–06 years/°C	11.89		
K374+975	Cement pavement	The maximum temperature	16.82	16.22	16.71
		The minimum temperature	−11.60	−12.44	−13.54
		The annual average ground temperature	14.21	14.33	15.12
		The average temperature amplitude of 04–06 years/°C	14.55		

K375+300	Asphalt pavement	The maximum temperature	19.17	19.79	19.32
		The minimum temperature	−11.02	−10.17	−11.41
		The annual average ground temperature	15.09	14.98	15.36
		The average temperature amplitude of 04–06 years/°C	15.15		

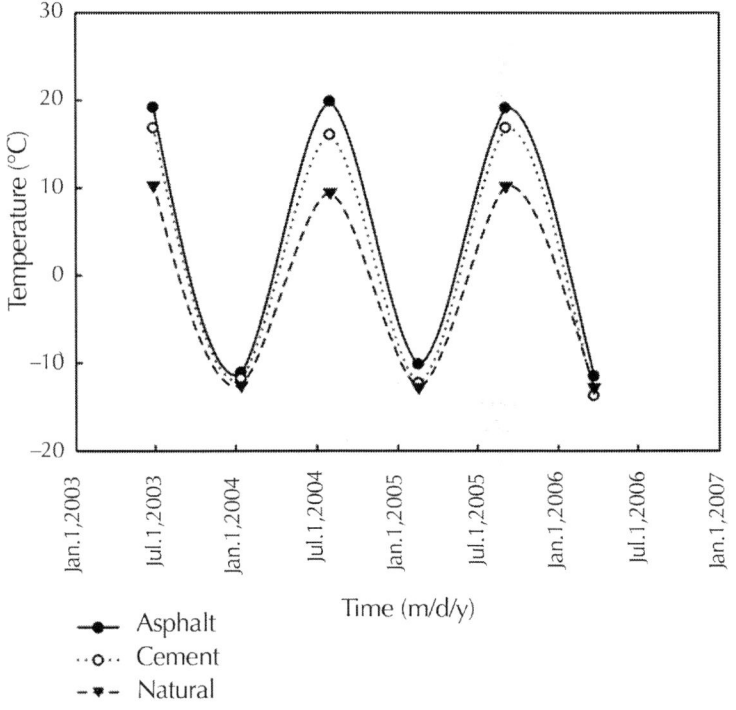

Figure 2: The curve of the surface temperature amplitude under different pavements.

FINITE ELEMENT ANALYSIS MODEL

Mathematical Model

The freezing and thawing cycle of permafrost embankments is the process of heat and mass transfer accompanying heat flow and redistribution so the mechanism of the internal water and heat function of the embankment in the permafrost regions may be considered to be a heat transfer considering moisture migration accompanying phase changes.

The freeze-thaw cycle in permafrost embankment is accompanied by the redistribution of temperature field and moisture migration. The mechanism of water and thermal in the permafrost embankment can be attributed to the thermal conduction problems accompanying phase changes. The following hypotheses are presented for the selected mathematical model considering the phase changes of freeze-thaw which influences the temperature field and seepage field of the permafrost embankment.

- Layers soil of each embankment section is homogeneous.
- No external load acts on soil layer during freezing and thawing.

Because thermal conduction term is far greater than convection term in the freeze-thaw process of frozen soil, the effect of the convection, mass transfer, the latent heat of vaporization, and chemical potential; are negligible in calculating analysis compared to the heat diffusion and the heat diffusion equation only considers soil skeleton and thermal conductivity of water, so the ice-water phase change may be written as follows

$$\frac{\partial}{\partial_x}\left(K_x \frac{\partial T}{\partial x}\right) + \frac{\partial}{\partial_y}\left(K_y \frac{\partial T}{\partial y}\right) = C\rho \frac{\partial T}{\partial x} - L\rho_i \frac{\partial W_i}{\partial t}, \qquad (1)$$

where T is transient temperature and t is time. K_x and K_y are components of soil equivalent thermal conductivity. C, r, L, W, and W_i are soil equivalent volumetric heat capacity, soil density, latent heat freezing and thawing of water, soil moisture, and unfrozen soil moisture, respectively. And x and y represent horizontal direction and vertical direction respectively in the cross section of the embankment.

Computational Region

Figure 3 shows the computational region. Three embankments widths of 8.5 m, 12.0 m, and 22.5 m, representing the highways with different grades (People's Republic of China Profession Standards, 1998), were selected. The gradient of the embankment slope is 1 : 1.5. The embankment height is adjustable ranging from 1 m to 3 m with a step of 1 m. The flank fields of both sides are 20 m from the foot of slope, and the lower boundary is 30 m below the natural ground surface. The thermal stability of permafrost embankment under asphalt concrete pavement and cement concrete pavement is analyzed with different heights under every embankment width.

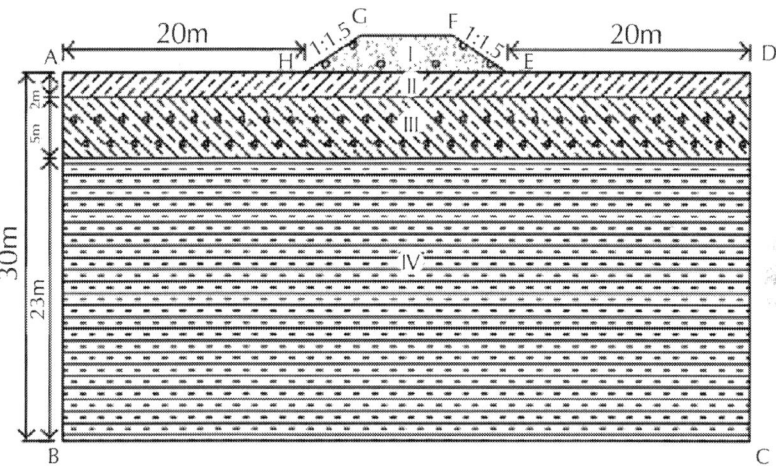

Figure 3: The computational region. Zone I is gravel backfill, zone II is sub-clay, zone III is crushed rock and sub-clay, and zone IV is argillaceous rocks.

The most major difference that permafrost has from other soils is that its property has close relationship with temperature. The heat capacity of the frozen soil skeleton only considers the volumetric heat capacity of freezing mode and thawing mode in computation. In addition, the thermal conductivity value only considers the effect of freeze-thaw state while ignoring the effect of temperature. Soil parameters within computational region are listed in Table 5 [3, 10].

Table 5: Soil parameters used in finite element analysis

The thermal parameters	l_f/(kJ/ (m · °C · d))	l_u/(kJ/ (m · °C · d))	C_f/(kJ/ (m³ · °C))	C_u/(kJ/ (m³ · °C))	Volumetric water content
Gravel backfill	129.60	120.96	1 827	2 226	0.08
Subclay	155.52	129.60	2 066	2 718	0.30
Rock and subclay	190.00	86.40	2 468	3 806	0.50
Argillaceous rocks	228.40	160.10	2 594	3 892	0.15

l_f and l_u are the thermal conductivity coefficients of frozen soil and unfrozen soil, respectively. C_f and C_u are the volumetric heat capacity coefficients of frozen soil and unfrozen soil, respectively.

The Boundary Conditions

The lower boundary temperature condition is determined by the secular measured ground temperature gradient at the depth of 30 m in Plateau permafrost region. The temperature gradient can be described as follows:

$$\frac{\partial T}{\partial y} = 0.02\frac{°C}{m} \qquad (2)$$

The temperature gradient of the temperature boundary condition of embankment is 0 in the horizontal direction due to the lateral natural ground of embankment away from the embankment. Consider the following equation:

$$\frac{\partial T}{\partial x} = 0 \qquad\qquad (3)$$

As the temperature boundary condition values of embankment slope are slightly lower than the upper temperature boundary condition values of cement concrete pavement, the upper temperature boundary condition of cement concrete pavement is simplified for the following trigonometric functions:

$$T = T_0 + R_0 t + A\sin\left(\frac{2\pi t}{365} + B\right) \qquad (4)$$

where T_0 is the initial annual ground temperature distribution of the embankment surface, T is operating time, is temperature amplitude of the embankment surface, R_0 is increasing rate of ground surface temperature caused by the global climate warming, $R_0 = 0.02°C/a$, B is the initial calculated phase, and A and T_0 are obtained by analyzing the measured temperature of Zuimatan testing segment of the national highway 214 in Table 6.

Table 6: The annual ground temperature and the ground temperature amplitude in the top boundaries

Pavement type	$T_0/°C$	The ground temperature amplitude $A/°C$
Natural ground surface/°C	−1.0	11.89
Asphalt pavement/°C	3.0	15.15
Cement pavement/°C	1.0	14.55
The gravel slope/°C	0.2	14.2

COMPUTATIONAL RESULTS AND ANALYSIS

Modeling Verification

According to the aforementioned boundary conditions, the initial conditions, and the thermal physical parameter of the unstable high temperature permafrost with an average annual temperature of −3.5°C, GEO-SLOPE is used separately to calculate the maximum thawed depth under the center line of cement concrete pavement in Figure 4 and asphalt pavement in Figure 5 one year after embankment constructed. As can be seen in Figures 2 and 3, the ground temperature curves separately calculated by the GEO-SLOPE are basically consistent with the measured temperature curve under cement concrete pavement and asphalt pavement respectively. Compared respectively to the measured temperature values of the K422+820 section of Zuimatan testing segment along the national highway 214 for cement concrete pavement and asphalt pavement, the maximum thawed depth under the permafrost embankment happened in November of 2004, that is when the 2nd year after Zuimatan testing segment construction of the national highway 214 was completed. The consistency between the calculated temperature curve and the measured temperature curve verifies the reliability of the mathematical model simultaneously.

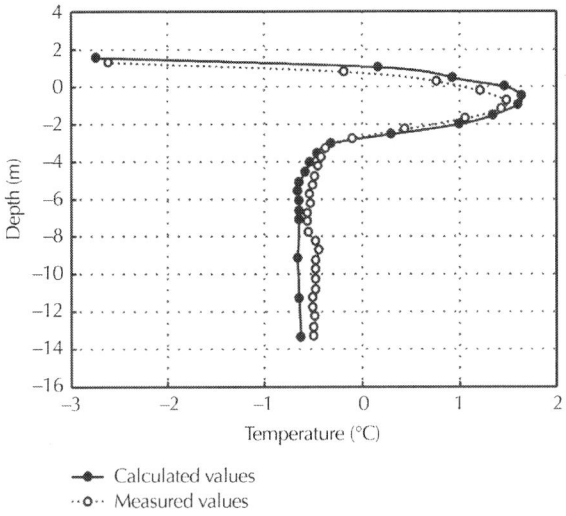

Figure 4: Measured and calculated temperature values at the maximum thawed depth of embankment center (1 year after the construction of cement concrete pavement).

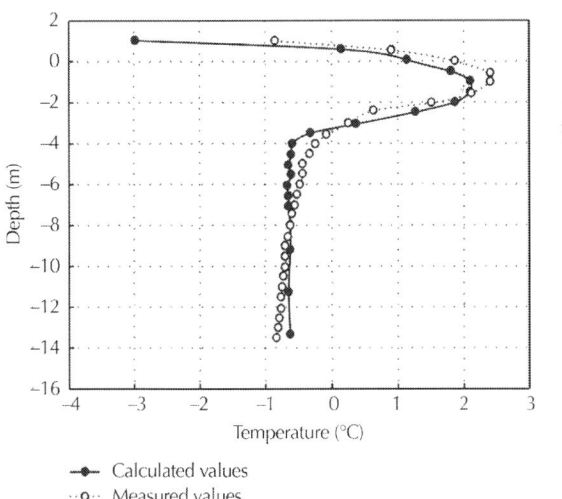

Figure 5: Measured and calculated temperature values at the maximum thawed depth of embankment center a year after the construction.

Results Analysis

The Thawing Core Generating. The temperature field under cement concrete pavement embankment and asphalt pavement embankment is firstly analyzed separately at a pavement width of 8.5 m. As can be seen in Figure 6, the residual thawed layers appeared under the embankments with a height of 1.5 m, 2.0 m, 2.5 m, and 3.0 m, respectively. The reason is that the temperature rises gradually after the two embankments constructed completely. The maximum thawed depth under permafrost embankments when the residual layers begin to appear is considered as the maximum thawed depth of permafrost embankments. The appearance time of the residual thawed layers has close relationship with pavement type, pavement height, and pavement width. Figure 6also shows the maximum thawed depth calculated under the two pavement structures 1a, 5a, 10a, 20a, 30a, 40a, and 50a (a representing year), respectively, after they were completely built. The 50th year temperature field under the two pavement structures is also seen from Figure 6. For the pavement width of 8.5 m, the maximum thawed depth under asphalt pavement has been always greater than cement concrete pavement every year from the first year to the 30th year after the embankments were completely built. The maximum thawed depth differences between the two pavements become greater and greater as the embankment height decreases.

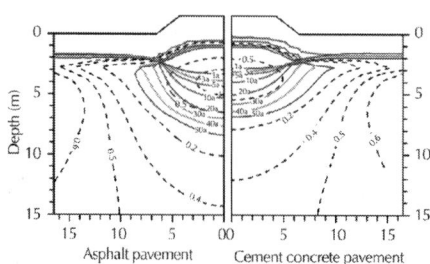

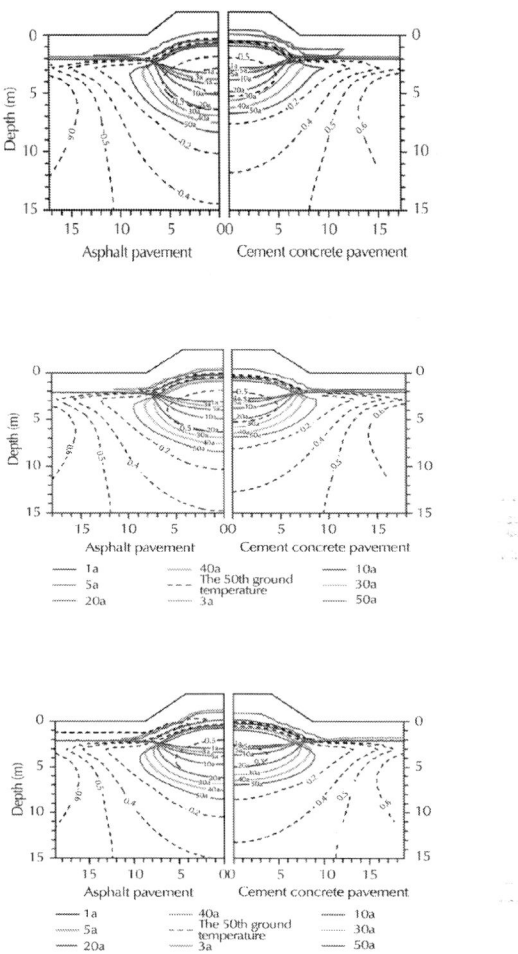

Figure 6: Variations of the maximum thawed depth and temperature value of the 50th year under concrete pavement and asphalt pavement (width of 8.5 m). (a) An embankment with a height of 1.5 m, (b) An embankment with a height of 2.0 m, (c) An embankment with a height of 2.5 m, and (d) An embankment with a height of 3 m.

Figure 6 shows that for asphalt pavement permafrost embankment and cement concrete pavement embankment, the maximum thawed depth differences between the both pavements structures present on decreasing tendency at the same embankment width as the embankment height increases. And the maximum thawed depth of the cement concrete pavement embankment is always less than asphalt pavement permafrost embankment at the same service time. Embankment height and embankment thermal resistance are considered as the main factors resulting in the maximum thawed depth differences. The external factors affect the temperature field under permafrost embankment through the pavement. The embankment thermal resistance is increasing gradually with the increase of the embankment height. The external factors influence less and less the temperature field of permafrost embankment through the pavement as the embankment height increases. The different material performances of pavement structures will generate different thermal resistances. Therefore, the degree of the external factors affecting the temperature field under permafrost embankment through the pavement will present differences because of the pavement material performances. For example, when the embankment height is 1.5 m, the 50th year maximum thawed depth under cement concrete pavement after embankment begins to operate is basically as much as the 10th year maximum thawed depth under asphalt pavement. However, the maximum thawed depth differences between both pavements become little as the embankment height increases although the 50th year maximum thawed depth under cement the concrete pavement after embankment begins to operate is greater than the 30th year one under the asphalt pavement. It can be concluded from the previous analysis that the thermal stability of permafrost under cement concrete pavement is obviously better than asphalt pavement at the same service time if taking the same lower embankment height.

It is found from Figure 6 that for cement concrete pavement and asphalt pavement, respectively, the time of the maximum thawed depth developing the fastest under every pavement appears in the first 20 years after the highways begin to operate at different embankment height. The developing rate of the maximum thawed depth under every pavement slows down significantly after 20 years of operation. The change is mainly caused by the instability of the temperature field under permafrost embankments disturbed by human activities and engineering construction after the highway embankment has been built. The temperature field of permafrost embankments begins to become extremely unstable when the maximum thawed depth increases. But after a long operation, the degree that the maximum thawed depth under every pavement is influenced little by external temperature, climatic conditions, and engineering construction gradually decreases. In this case, the temperature field under permafrost embankment tends to stabilize. And the developing rate of the maximum thawed depth shows the characteristics of first increasing and then decreasing with the embankment operation time increasing. In addition, the time when the thawing core appears under asphalt pavement permafrost embankment is significantly earlier than cement concrete pavement during the operation time when the embankment height is 1.5 m, 2.0 m, 2.5 m, and 3.0 m, respectively. To every embankment, the range of pavement type affecting the temperature field under permafrost embankment is approximately within 5 m distant from outside embankment slope when the width of the two embankments is 8.5 m. The range and degree of the pavement material types influencing the temperature field of permafrost embankment are closely related to the height of embankment. But in general, the range of pavement type influencing the temperature field under permafrost embankment presents a decreasing trend with the increment of the embankment height.

To the permafrost embankment with a width of 8.5 m, the range of pavement type influencing on the temperature field under permafrost embankment is approximately within 5 m distant from outside embankment slope. And the pavement types have no influences on the temperature field under permafrost embankment outside 5 m distant from outside embankment slope. The range and degree that the pavement types influence on the temperature field under permafrost embankment have close relations with the height of embankment. But in general, the range of pavement type influence on the temperature field under permafrost embankment presents decreasing trend with the increment of the embankment height.

From the analysis of the previously calculated results, we can find that the maximum thawed depth under the cement concrete pavement permafrost embankment has increased not obviously with the highway permafrost embankment service time increasing compared to the asphalt pavement although the maximum thawed depth of cement concrete pavement has increased with the highway permafrost embankment service time increasing.

In order to clearly describe that the thermal stability of permafrost under cement concrete pavement is better than that under asphalt pavement, Figure 7 shows the relationship between the maximum thawed depth under the cement concrete pavement and service time and the different embankment height. The maximum thawed depth differences under the asphalt pavement and cement concrete pavement under different pavement widths in the high ice content permafrost region are with an average temperature of −3.5°C. The relationship between the maximum thawed depth under the cement concrete pavement and the service time varies because of the different width of cement concrete pavement and asphalt pavement. For example, the maximum thawed depth differences under both pavements with a pavement width of 8.5 m increase with the increment of service time and then tend to decrease a certain time after operation. But when the pavement width of the two pavements is 12 m and 22.5 m, respectively, the relationship between the maximum thawed depth differences under the two pavements and service time approximately present on nolinear increasing. It also shows that the pavement width greatly has influences on the change rate of the maximum thawed depth difference. The maximum thawed depth differences under two pavements increase apparently with the increment of service time when the pavement width of the embankment becomes larger. It also shows that the larger width of the cement concrete pavement is an advantage to the thermal stability under the cement concrete pavement.

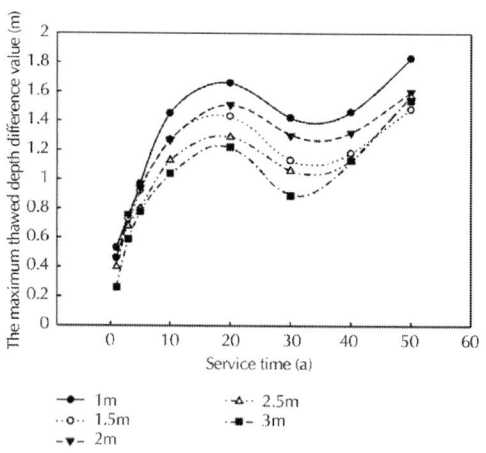

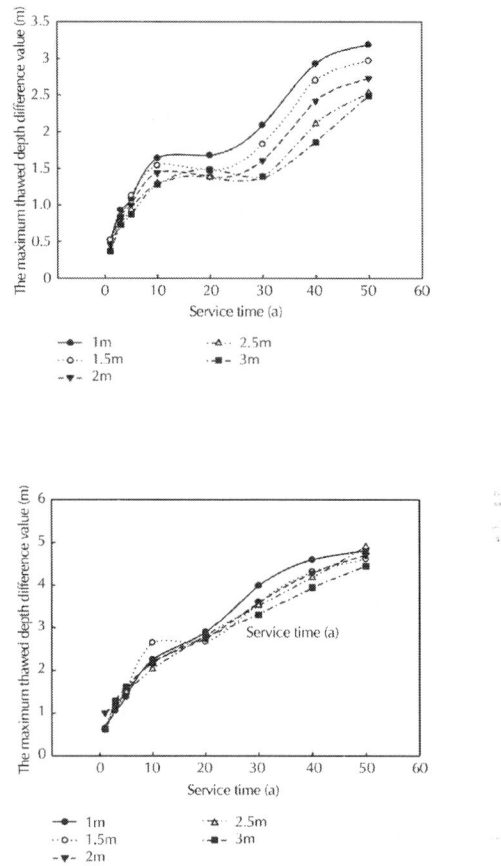

Figure 7: The maximum thawed depth difference value under concrete pavements and asphalt pavements versus service time curve under different pavement width. (a) A pavement with a width of 8.5 m, (b) A pavement with a width of 12.0 m, and (c) A pavement with a width of 22.5 m.

The Temperature Differences under Different Pavement Types

The temperature field under every pavement with the same embankment height presents different characteristics under different time or different embankment width. We can find that the 50th year maximum thawed depth under each pavement changes with different reasons in Figures 8 and 9 when the embankment width is 8.5 m.

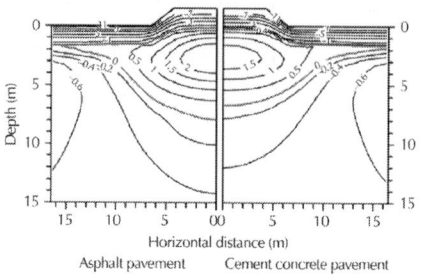

Asphalt pavement   Cement concrete pavement

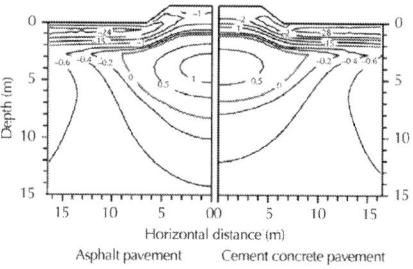

Asphalt pavement   Cement concrete pavement

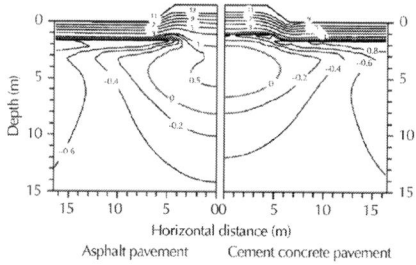

Asphalt pavement   Cement concrete pavement

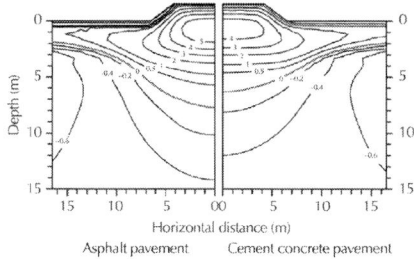

Figure 8: Temperature values of the 50th year under concrete pavement and asphalt pavement (width of 8.5 m). (a) The ground temperature field on January 10, 2053, (b) The ground temperature field on March 30, 2053, (c) The ground temperature field on July 20, 2053, and (d) The ground temperature field on November 10, 2053.

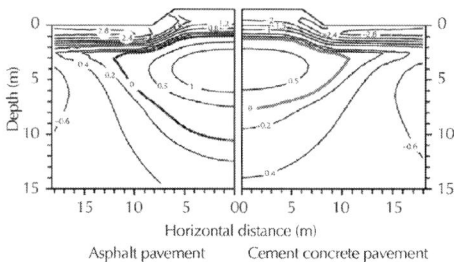

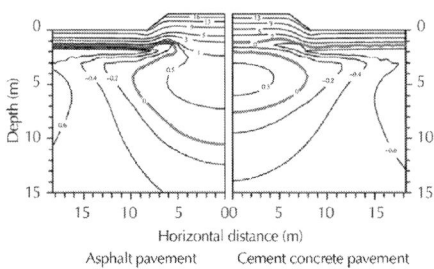

Figure 9: Temperature values of the 50th year under concrete pavement and asphalt pavement (width of 12.0 m). (a) The ground temperature field on March 30, 2053 and (b) The ground temperature field on July 20, 2053.

Figure 8 shows the 50th year ground temperature field under both pavements with a width of 1.5 m after the embankment service. It is seen from Figure 8 that January 10 is the lowest temperature, March 30 is the maximum thawed depth, July 20 is the maximum air temperature, and November 10 is in freezing period. In freezing period (January 10), there exists a close thawed core that is higher than 0°C under each pavement. The ground temperature curves are parallel to the embankment surface profile above the close thawed core. The ground temperature and ground temperature gradients between pavement and plane at 1.5 m depth are both apparently higher than the ones between natural ground and plane at 1.5 m depth. On the contrary, the ground temperature curves have large differences under the close thawed core. The differences are induced by pavement material performances. The shallow ground temperature under every pavement begins to rise when the maximum thawed depth appears under both embankments On March 30, 2053. The differences between the shallow ground temperatures also begin to increase gradually. Comprised with the ground temperature field in freezing period, the ground temperature field differences are not apparent. When the ground temperature reaches the highest temperature on July 20, 2053, the isothermal curve of the shallow ground is basically parallel to the embankment surface profile. But the shallow ground temperature under the asphalt pavement is higher 2.0°C than that of the concrete pavement. There is a more semienclosed thawed zone under the asphalt pavement. And there is a close thawed zone (thawed core) under the concrete pavement. The ground temperature in the thawed core of the concrete pavement is lower than that of the asphalt pavement. This is mainly caused by the concrete pavement having stronger temperature susceptibility than the black asphalt pavement. From November 10, the shallow ground temperature under every pavement begins to decrease. The maximum thawed depth in the embankment center is far greater than the natural ground. But the maximum thawed depth in the embankment center is still less than the artificial one. Although the natural ground begins to freeze, the time of the asphalt pavement refreezing is apparently later than that of the concrete pavement.

Figures 9 and 10 show that for the embankments with widths of 12.0 m and 21.5 m, there are big differences at the ground temperature under pavements. The ground temperature curves are basically identical for the embankments with widths of 8.5 m, 12.0 m, and 21.5 m. This indicates that pavement materials influence the difference distribution. Particularly in summer, the shallow ground temperature differences are very obvious between the concrete pavement and the asphalt pavement. There are small shallow ground temperature differences between the concrete pavement and the asphalt pavement. This proves further that the concrete pavement has stronger temperature susceptibility than the black asphalt pavement.

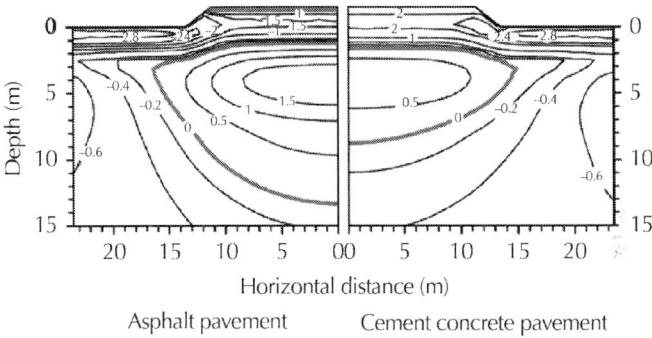

Figure 10: Temperature values of the 50th year under concrete pavement and asphalt pavement (width of 21.5 m). (a) The ground temperature field on March 30, 2053 and (b) The ground temperature field on July 20, 2053.

It is seen from Figures 9 and 10 that in different seasons, pavement materials greatly have influence on the ground temperature under the embankment. On the contrary, pavement widths influence less the ground temperature under the embankment. Therefore, the ground temperature change under the embankment by the pavement material is more than that by the pavement widths. In addition, the influence area of the temperature field under both the asphalt pavement and the concrete pavement is distant 5 m or so from the embankment slope. Outside 5 m distance from the embankment slope, the influence area of the temperature field under both the asphalt pavement and the concrete pavement becomes less. The influence area has begun to decrease with the increment of embankment height.

CONCLUSIONS

From the aforementioned results and analyses, we can find several significant conclusions for the thermal stability analysis under embankment with asphalt pavement and cement pavement in permafrost region.(1)There is a ground temperature difference between asphalt pavement and concrete pavement at the same depth within 8 m under pavement. The ground temperature difference changes with the seasons. The difference is 2.2–5.2°C in summer while the difference is basically less than 1°C in winter.(2)The maximum thawed depth difference between asphalt pavement and cement pavement decreases with the increment of the embankment height. The development rate of the maximum thawed depth begins to increase with the embankment height decreasing. The time when the development rate of the maximum thawed depth appears is the first 20 years after the embankments operate. From then on, the development rate of the maximum thawed depth begins to decrease.(3)The temperature under cement concrete pavement is always lower than that of asphalt pavement at the same service time in the computational region. To the permafrost embankment with the same width, the maximum thawed depth under asphalt pavement is greater than that of cement concrete pavement. The maximum thawed depth difference between the two pavement structures becomes more significant with the embankment width increasing.(4)The developing rate of the maximum thawed depth presents the characteristic of first increasing and then decreasing under different embankment height. The appearance time of the thawing core under asphalt pavement is earlier than that of cement concrete pavement.(5)In different seasons, pavement materials greatly influence the ground temperature under the embankment. On the contrary, pavement widths influence less the ground temperature under the embankment. Therefore, the ground temperature change under the embankment by the pavement material is more than that by the pavement widths. In addition, the influence area of the temperature field under both the asphalt pavement and the concrete pavement is distant 5 m or so from the embankment slope. Outside 5 m distance from the embankment slope, the influence area of the temperature field under both the asphalt pavement, and the concrete pavement becomes less. The influence area has begun to decrease with the increment of embankment height.(6)The maximum thawed depth differences under the two pavement structures present the trend of decreasing with the service time increasing a certain time after the pavements begin to operate if the pavement widths are less. The maximum thawed depth differences between the two pavement structures approximately present linearity with operating time if the pavement widths are greater. The maximum thawed depth under cement concrete pavement is always less than asphalt pavement.(7)It is more important to assure the sustainability of the embankment engineering in the high temperature permafrost region, and the effective measures must be taken to protect permafrost because of the extremely fragile ecological environment.

ACKNOWLEDGMENTS

This paper was supported by the Fundamental Research Funds for Department of Education of Sichuan Province, no. 13ZA0188, by the Fundamental Research Funds for the Central Universities, no. SWJTU12CX063, and by the National Natural Science Foundation of China (no. 51208435).

REFERENCES

1. Y. Lai, L. Zhang, S. Zhang, and J. Xiao, "Adjusting temperature distribution under the north and south slopes of roadbed by the ripped-rock revetment in permafrost regions," Chinese Journal of Rock Mechanics and Engineering, vol. 23, no. 24, pp. 4212–4220, 2004.

2. Y. Sheng, W. Ma, Z. Wen, and M.-Y. Zhang, "Analysis of difference in thermal state between south faced slope and north faced slope of railway embankment in permafrost region," Chinese Journal of Rock Mechanics and Engineering, vol. 24, no. 17, pp. 3197–3201, 2005.

3. J. Li and Y. Sheng, "Analysis of the thermal stability of an embankment under different pavement types in high temperature permafrost regions," Cold Regions Science and Technology, vol. 54, no. 2, pp. 120–123, 2008.

4. Y. Chou, Y. Sheng, and Z. Wei, "Evaluation on thermal stability of embankments with different strikes in permafrost regions," Cold

Regions Science and Technology, vol. 58, no. 3, pp. 151–157, 2009.

5. Q. H. Yu, Y. Z. Liu, and C. J. Tong, "Analysis of the subgrade deformation of the Qinghai-Tibetan Highway," Journal of Glaciology and Geocryology, vol. 24, no. 5, pp. 623–627, 2002.

6. S. L. Wang, L. Zhao, S. X. Li, et al., "Study on thermal balance of asphalt pavement and roadbed stability in permafrost regions of the Qinghai-Tibet Highway," Journal of Glaciology and Geocryology, vol. 23, no. 2, pp. 111–117, 2001.

7. Y. Z. Liu, Q. B. Wu, J. M. Zhang, and Y. Sheng, "Deformation of highway embankment in permafrost regions of the tibetan plateau," Journal of Glaciology and Geocryology, vol. 24, no. 2, pp. 10–15, 2002.

8. J. Pei, M. Dou, C. Hu, Z. He, and W. Zhang, "Application of geosynthetics for treatment of embankment troubles in permafrost regions," Journal of Glaciology and Geocryology, vol. 24, no. 2, pp. 785–789, 2002.

9. M. Dou, C. Hu., Z. He, et al., "Distributing regularities of embankment diseases in permafrost section of the Qinghai-Tibetan Highway," Journal of Glaciology and Geocryology, vol. 24, no. 6, pp. 780–784, 2002.

10. J.-P. Li and Y. Sheng, "Study on diseases of cement concrete pavement in permafrost regions," Cold Regions Science and Technology, vol. 60, no. 1, pp. 57–62, 2010.

11. G. Yong, D. Li, K. Zhang, et al., "Investigation and analysis of road pavement on asphalt and concrete,"Highway, no. 12, pp. 158–163, 2009.

12. J. W. Zhang, J. P. Li, and X. J. Quan, "Disease mechanism of embankment with asphalt pavement and cement pavement in permafrost regions," Disaster Advances, vol. 6, supplement 2, pp. 18–32, 2013.

13. J. Zhang, Z. Li, and J. Wu, "Major influencing factors analyses of stability of permafrost embankment,"Highway, no. 2, pp. 17–20, 2000.

14. Y. Chou, Y. Sheng, Y. Li, Z. Wei, Y. Zhu, and J. Li, "Sunny-shady slope effect on the thermal and deformation stability of the

highway embankment in warm permafrost regions," Cold Regions Science and Technology, vol. 63, no. 1-2, pp. 78–86, 2010.

15. F. Yu, J. L. Qi, X. L. Yao, and Y. Z. Liu, "Degradation process of permafrost underneath embankments along QinghaiTibet Highway: an engineering view," Cold Regions Science and Technology, vol. 85, no. 1, pp. 150–156, 2013.

16. H. Liu, F. Niu, Y. Niu, Z. Lin, J. Lu, and J. Luo, "Experimental and numerical investigation on temperature characteristics of high-speed railway's embankment in seasonal frozen regions," Cold Regions Science and Technology, vol. 81, pp. 55–64, 2012.

17. L. Jin, S. Wang, J. Chen, and Y. Dong, "Study on the height effect of highway embankments in permafrost regions," Cold Regions Science and Technology, vol. 84, pp. 122–130, 2012.

Chapter 9

Effects of Crumb Rubber Size and Concentration on Performance of Porous Asphalt Mixtures

Altan Cetin

Civil Engineering Department, Engineering Faculty, Anadolu University, 26555 Eskisehir, Turkey

ABSTRACT

The purpose of this study is to investigate the effect of size distribution and concentration of crumb rubber on the performance characteristics of porous asphalt mixture. The recycling of scrap tires in asphalt pavements appears as an important alternative providing a large-scale market. The characteristics of bitumen are very important with regard to service life of porous asphalt pavement. The experimental study consists of two main steps. Firstly, the mixture design was performed to determine the optimum bitumen content. In the latter step, the mixtures were modified by dry process using crumb rubber in three different grain size distributions of #4~#20, #20~#200, and #4~#200 and rubber content of 10%, 15%, and 20% as weight of optimum bitumen. The permeability, Cantabro abrasion loss, indirect tensile strength, moisture susceptibility, and resilient modulus tests were carried out on the specimens. Test results show that #20~#200 sized rubber particles reduced air voids and coefficient of permeability, while they increased the Cantabro abrasion loss. In general, increasing the crumb rubber size and content decreased the performance characteristics of the porous asphalt mixtures.

INTRODUCTION

The worn-out tires from vehicles leave billions of waste tires every year becoming a significant source of waste materials. Scrap tires are still a serious environmental and financial issue for many countries in the world occupying landfill spaces and becoming a threat for health and safety hazards to the community. The scrap tires consist of rubber, carbon black, steel, and so forth potentially to be very useful in various applications which have been evaluated effectively as a valuable resource. There are some different recycling strategies developed for waste tires. The main markets in the assessment of waste tires include tire-derived fuel (TDF), civil engineering applications, and ground rubber. The recycling of waste tire has a good trend. However, it has still potential to consume more waste tires. In addition, TDF corresponds to 54% of the total scrap tires which may not represent an ideal application in the current recycling methods for waste tires from environmental conservation perspective [1]. More value-added alternatives are required to be discovered in order to motivate public and private agency to recycle them.

One of the approaches of recycling scrap tires is to use crumb rubber from the tires as a component in asphalt pavement mixtures. The crumb rubber is combined in asphalt mixtures to improve the performance of asphalt concrete pavements. The large-scale usage of crumb rubber from waste tyres in asphalt mixtures appears to be more feasible alternative in terms of engineering applications and environmental consideration. Asphalt-rubber pavements can minimize environmental impact and maximize the conservation of natural resources. The ability of crumb rubber to improve the asphalt mixture performance depends on many factors such as the mixing methods, reaction time with bitumen, nature of the rubber, and size and concentration of the rubber particles. There are two different processes using crumb rubber in the asphalt mixtures, a dry process and a wet process. In the wet process, the finer crumb rubber is mixed with asphalt cement at high temperature. It reacts with the bitumen and creates modified bitumen. In the dry process, the crumb rubber is mixed together with the aggregates prior to the addition to the asphalt. During dry process in which the crumb rubber is used as an aggregate, the chemical reaction between crumb rubber and bitumen is quite limited [2].

The advancement in hot-mix asphalt pavement technology resulted in the development of a different type of asphalt mixture. Porous asphalt is used widely as an application of pavement surface in Europe. Porous asphalt or open-graded asphalt concrete is an environmentally friendly road material which was developed by using advanced technology in hot-mix asphalt mixture design. It is used effectively in regions with the highest level of precipitation. The porous asphalt used as surface layer mostly has an air void content of 20% and can improve the ride quality for drivers during wet weather by reducing skidding. This pavement system prevents hydroplaning and spraying on the road surface and improves visibility by eliminating the light reflected from the road surface. The porous asphalt mixture is designed by using a relatively large proportion of coarse aggregates (more than 80%) with few fine aggregates to create a large space for water drainage. It also significantly reduces traffic induced noise emissions by means of the high porosity [3]. The selection of asphalt binder with high viscosity is an important issue on account of the specific structure of its high void content in the mixtures. To improve the durability of porous asphalt mixtures, polymer-modified bitumen was recommended to be used in moderate and hot climates with heavy traffic [4].

Modified bitumen with polymers and synthetic and natural rubber used in porous asphalt surfaces significantly reduce the level of noise from vehicles compared with the dense graded asphalt surfaces [5]. The use of polymer-modified or rubberized binders instead of unmodified binder in the 4.75 mm open-graded mixture reduced permeability but increased acoustic absorption. Mixtures containing rubberized bitumen show the most acoustic absorption improvement [6]. One advantage of using rubber modified asphalt is that it raises the viscosity of the bitumen and provides increase of the bitumen content in the mixture. Thus, massive asphalt film surrounding the aggregate enhances durability of the asphalt mixtures. The open-graded asphalt with asphalt-rubber binder had significantly reduced not only moisture sensitivity but also superior fatigue resistance and cracking as compared to traditional porous or semiporous asphalt mixtures [7, 8]. A previous study claim that Large Stone Porous Asphalt-Rubber Mixture has better performance than conventional large stone porous asphalt mixture using polymer-modified asphalt, and it could be used as stress absorbing layers of semirigid base asphalt pavement [9]. Moreover, the experimental results of repeated triaxial-loaded and wheel tracking permanent deformation tests confirmed that the asphalt-rubber porous mixtures have superior performance against rutting [10]. Several studies have shown that rubber content between 10% and 30% of the bitumen improves resistance to moisture damage, the susceptibility to temperature, and the tendency to flow [11, 12]. The particular facts of performance evaluation of test sites in Florida indicated that the wet process addition of asphalt-rubber blended into the bitumen improved the cracking resistance of the surface mixtures [13]. The characteristics of the crumb rubber can influence properties of asphalt rubber such as rubber quantity in the blend and particle size distribution. Additional factors include crumb rubber surface area, grinding process, crumb rubber chemical composition, and contaminants as water, fibre, and metal [14].

Considering the preceding information from the literature, this study was designed to examine the effects of the content and size of crumb rubber on the porous asphalt mixture performance.

EXPERIMENTAL PROGRAM

This section includes material characterization of basalt aggregate, bitumen and crumb rubber, and test methods such as permeability, Cantabro abrasion loss, indirect tensile strength, moisture susceptibility, and resilient modulus tests.

Materials Characterization

Crushed basalt aggregates and conventional bitumen of penetration grade 50/70 were used in this experimental study. Also, crumb rubber obtained from waste tires was used as a modifier in different sizes and concentrations. The characterization tests conducted in the laboratory were given in this section.

Aggregate

High-quality aggregates are required to provide the field performance of the large-void porous asphalt during the service life. In this study, one type of crushed basalt aggregate was used as coarse and fine aggregate. Also, stone dust which was obtained from the same basalt aggregate was used as the filler material. The basalt aggregate was provided from a quarry in Eskişehir, Turkey. The main physical properties of the coarse aggregate with the criteria are given in Table 1. The aggregate specific gravity and absorption test results were shown in Table 2.

Table 1: The physical properties of coarse aggregate

Properties	Specification	Results	Specification	
			Min.	Max.
Los Angeles abrasion (%)	ASTM C131	13.5	—	30
Sodium sulfate soundness (%)	ASTM C88	0.64	—	12
Percent fractured faces (%)	ASTM D5821	100	100	—
Flakiness index (%)	BS 812	20	—	25
Polish value	ASTM C3319	53	50	—
Stripping resistance (%)	ASTM D1664	60–65	—	50

Table 2: Test results on aggregate specific gravity and absorption properties

Properties	Course aggregate	Fine aggregate	Filler
Bulk specific gravity	2.586	2.639	—
Apparent specific gravity	2.735	2.763	2.782
Water absorption (%)	2.1	1.7	—

The aggregate gradation for porous asphalt mixtures was selected between broad bands (boundary lines) according to Porous European Mix (PEM) Specification. The aggregate gradation is given in Figure 1.

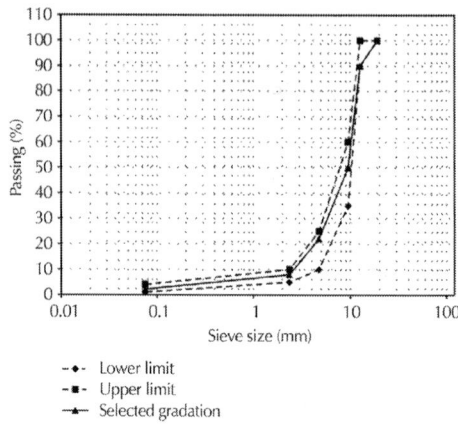

Figure 1: Aggregate gradation of porous asphalt mixture with specification limits of PEM.

Bitumen

Bitumen from one source was used in this study. The penetration grade bitumen of 50/70 obtained from the Asphalt Production Refinery is widely used in Turkey. This type of asphalt binder was chosen instead of modified bitumen so that effect of the crumb rubber on porous asphalt mixtures could be clearly determined. Table 3 gives physical properties of the bitumen.

Table 3: Physical properties of conventional bitumen

Properties	Specification	Results
Penetration at 25°C, 100 g, 5 s (0.1 mm)	ASTM D5	63
Softening point (°C)	ASTM D36	49
TFOT residue	ASTM D2872-04	
Mass loss (%)	ASTM D2872	0.096
Retained penetration (%)	ASTM D5	37
Softening point after hardening (°C)	ASTM D36	61

Flashing point (°C)	ASTM D92	240
Specific gravity at 25°C	ASTM D70	1.021

Crumbed Rubber

The crumb rubber was obtained from tire buffing process. It includes removing the worn tread by a special machine and applying a new tread. The different sizes of rubber were separated by sieves. The grain size distributions of crumb rubber were given in Figure 2. The grain size distribution group of #4~#20, #20~#200, and #4~#200 includes different sizes of rubber particles. Figure 3 represents a general appearance of each group. Even though Figure 3 gives some information about the form of #4~#20, the forms of small particles such as #100~#200 mesh size cannot be determined from it. Therefore, Scanning Electron Microscopy (SEM) images of the crumb rubber in Figure 4 were presented. The fiber-like shape of the pine needle rubber can contribute to reinforcing the porous pavement and decrease the bitumen drain down.

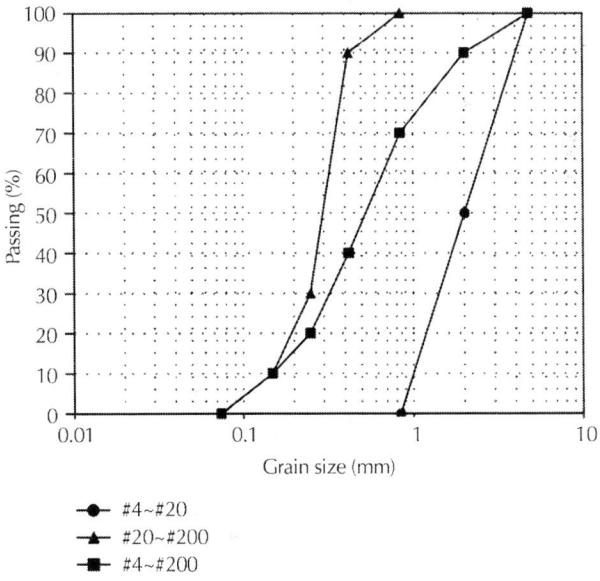

Figure 2: Grain size distributions of crumb rubber.

Figure 3: Appearance of crumb rubber particles: (a) #4~#20; (b) #20~#200; (c) #4~#200.

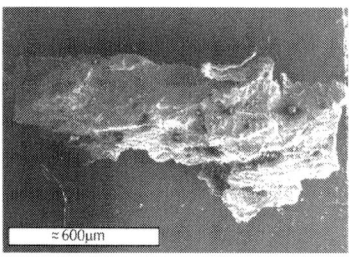

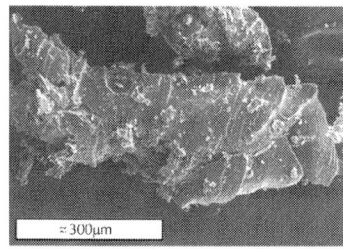

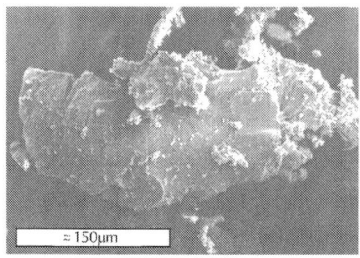

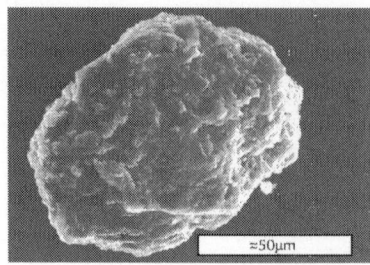

Figure 4: Scanning Electron Microscopy (SEM) pictures for different sizes of crumb rubber particles: (a) #4~#20; (b) #20~#200; (c) #4~#200; (d) #100~#200; (e) #100~#200.

Test Methods

The test program in this study consisted of two steps. Step I covered the mix design of porous asphalt mixtures. Step II was to investigate the effect of crumb rubber concentration and size on design and performance characteristics of the porous asphalt mixtures.

The principle design of porous asphalt mixtures is to determine the bitumen content that will optimize its engineering properties in relation with the in-service behavior during pavement life. The Porous European Mix Specification including the tests of air void, permeability, particle loss resistance (Cantabrian), and indirect tensile were used to determine the optimum bitumen content. In addition, performance-related tests such as moisture susceptibility and resilient modulus tests were conducted on the porous asphalt mixtures with and without crumb rubber.

Mixing and Production

Each specimen was comprised of about 1200 g of aggregate batches and bitumen. Modified specimens were prepared by dry process in which crumb rubber was added in the aggregate batch and mixed for 2 minutes. The aggregates and asphalt binder were blended at their corresponding mixing temperatures. A compaction process was followed subsequently by applying 50 blows on each face of the specimen using the standard Marshall hammer.

Air Voids Determination

To provide sufficient permeability in the porous asphalt mixtures, an air void content ranging between 16% and 22% (or greater) is recommended [15, 16]. The air void percentages of the porous asphalt specimens were very difficult to determine due to the higher porosity. This test procedure determines the bulk specific gravity of specimens of compacted asphalt mixtures. In this method, compacted specimens were coated with paraffin film, and then the bulk specific gravity with regard to the procedure in AASHTO T275 [17] was determined.

Hydraulic Conductivity Test

Hydraulic conductivity is a significant characteristic of porous asphalt mixtures on account of designing a drainage layer in pavement structures. The hydraulic conductivity of the compacted specimens symbolized in terms of the coefficient of permeability (k) was determined by using a falling-head water permeameter. The apparatus of falling-head permeability test consists of a metal cylinder and demountable metal plates at the top and bottom of the metal mold. The top plate has a hole with at least 31.75 mm inner diameter of graduated pipe for water inflow, and the bottom plate has an a outlet hole of minimum inner diameter of 18 mm and valve so that water can flow out. The specimen in the mold was placed between the bottom plate and the top plate and compressed by using clamps for sealing the bottom and top plates. The graduated pipe is filled with distilled water, and the valve is opened to flow through a saturated specimen. The time period taken for level change of water between two fixed points on the perspex pipe was recorded. The coefficient of permeability is then determined based on Darcy's law. A minimum permeability coefficient of 10^{-2} cm/s ($\approx$100 m/day) is commonly recommended for the pervious pavement structure [18, 19].

Cantabro Abrasion Test (Particle Loss Resistance)

The Cantabro test was conducted to evaluate the resistance to particle loss of the mixtures according to ASTM D7064 [20]. The compacted specimens were individually put in the Los Angeles testing machine without steel balls. After Los Angeles drum had been rotated for 300 revolutions at a speed of 30–33 revolutions per minute, the loose material broken off from surface of the test specimen was discarded. The masses of the specimens before and after the test were recorded. The percentage loss by weight of original specimen was calculated as the Cantabro abrasion. The percentage of Cantabro abrasion must be less than 25% in the European Specification.

Indirect Tensile Strength Test

The indirect tensile (IDT) strength test has been widely used for hot-mix asphalt (HMA) mixture design. The splitting strength is determined in this test as an indicator of the tensile strength of the compacted specimen. The results of (IDT) strength are employed to obtain the comparative relative strength of asphalt mixtures and predict the potential for pavement distress. Since the performance of porous mixtures depends on tensile strength of bitumen film, the IDT strength is also an important characterization test for porous asphalt mixture. The test is performed by using Marshall stability test equipment at 50 mm/min deformation rate and 25°C temperature in accordance with ASTM D6931 procedure [21]. In the test, a cylindrical specimen is exposed to a compressive load. It acts along the vertical diameter plane by a curved loading strip. The developed tensile stress, perpendicular to the direction of the applied load, ultimately causes the specimen to fail by splitting along the vertical diameter. The ultimate load at failure is recorded and used to calculate the IDT strength of the specimen.

Moisture Susceptibility

Moisture causes a loss of adhesion between the bitumen and the aggregate surface and accelerates the process of distresses such as rutting, cracking, and raveling in the asphalt mixture. Moisture susceptibility is extremely an important characteristic for the performance of the porous asphalt mixtures exposed to water damage. The test was performed according to Modified Lottman Test (AASHTO T283) that is one of the most largely used procedures for determining HMA water damage [22]. Six compacted specimens are required at 6%–8% air voids in the Modified Lottman Test. The specimens are separated into two groups. The first group of three is unconditioned specimens as the control group. The second group is conditioned specimens of vacuum-saturated saturation level between 55 percent and 80 percent. After conditioned specimens are placed in a freezer at −18°C for 16 hours, they are moved to at 60°C water bath for 24 hours. All of the specimens are subjected to the IDT strength test conducted with a loading rate of 50 mm/min at 25°C. The tensile strength ratio (TSR) is calculated by the average IDT strength of conditioned subset divided by the average IDT strength of control subset. The allowable value of TSR must be more than 70% for this test method.

Resilient Modulus (Stiffness Modulus) Test

Resilient modulus of asphalt mixtures is the most popular form of stress-strain measurement used to evaluate elastic properties. The five-pulse indirect tensile modulus test used to determine the stiffness of material was performed in accordance with ASTM D4123 [23] using Universal Testing Machine (UTM-5P). In the test, the cylindrical specimen is subjected to a pulsed diametric loading force, and the resulting total recoverable diametric strain is then measured at axes 90° from the applied force. Because the strain in the same axes is not measured, a value of 0.4 for Poisson's ratio of asphalt mixtures is accepted as a constant. The specimens were placed in the indirect tensile test equipment, and test results were recorded in the computer by data logger system. The test load sequence consists of 150 conditioning pulses and five-pulse test periods. The conditioning stage provides that the loading plates are seated onto the specimen for consistent results. The stiffness module is calculated by five-pulse test period. In addition, the specimen's skin and core temperatures were measured by transducers inserted in a dummy specimen located near the test specimen in order to control the testing temperature. In this study, specimens prepared in the laboratory were tested under the waveform type of Haversine at load pulse period of 3000 ms, pulse width of 80 ms, and peak loading force of 1000 N.

TEST RESULT AND DISCUSSION

The results of porous asphalt mixture design were given in this section. In addition, the effects of rubber size and concentration on performance characteristics were discussed.

MIX DESIGN

European Porous Asphalt Mix Design Approach is generally based on determining the voids and the percentage of particle abrasion loss at various binder contents. The design bitumen content is optimized for air voids and abrasion. It is recommended that the modified bitumen is used to improve the resistance against particle loss. It achieved a longer durability by means of its higher cohesion and viscosity [4].

Porous asphalt design procedures used in Europe and America mostly include the air void ratio, permeability, Cantabro abrasion loss, and indirect tensile strength tests. In this study, the following design criteria widely accepted in these countries were selected. The minimum permeability coefficient depends on the target air voids (18–23%) and is 100 m/day. A maximum Cantabro particle loss of 25% is allowed at 25°C. The allowable asphalt content ranges 4–6 percent. The bitumen content ensuring these limit values was selected as a percentage of optimum bitumen. The graphics of the mix design were given in Figure 5.

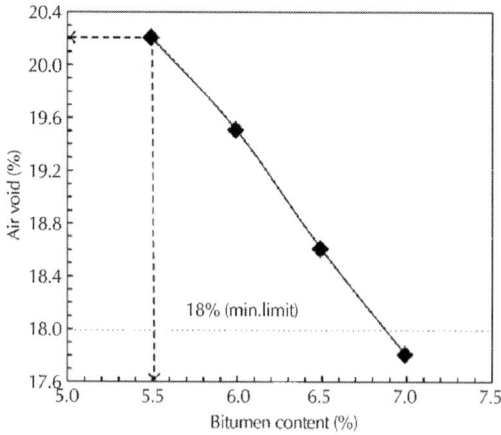

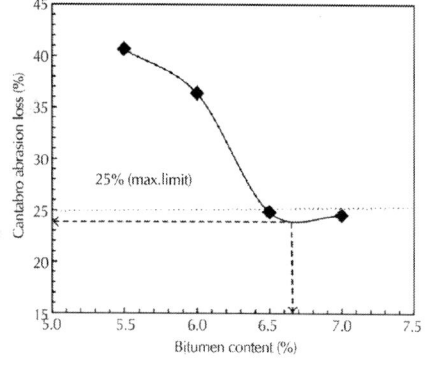

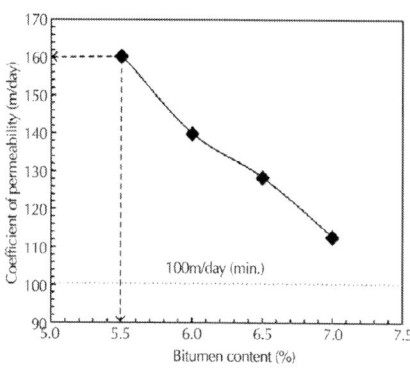

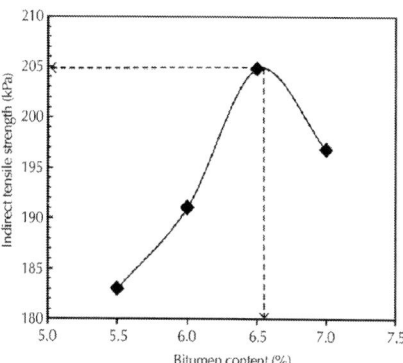

Figure 5: Graphics of mix design for porous asphalt mixtures: (a) air voids; (b) Cantabro abrasion loss; (c) coefficient of permeability; (d) indirect tensile strength.

The air void and permeability coefficients were decreased by increasing bitumen content (Figures 5(a) and5(c)). The maximum values of these characteristics were obtained for bitumen content of 5.5%. The minimum Cantabro abrasion loss provided the bitumen content of 6.5%. However, test result at bitumen content of 7% is quite close to results at 6.5% as shown in Figure 5(b). The maximum value for indirect tensile strength was also obtained for bitumen content of 6.5% as shown in Figure 5(d). The optimum bitumen content of 6.5% was determined according to these results. All design values of porous asphalt mixture were given in Table 4. Although the mix design of porous asphalt mixture was achieved, design requirements with the penetration grade bitumen of 50/70, the low values in Cantabro abrasion loss, and IDT strength test were introduced due to the necessity of modified bitumen.

Table 4: Result of the Porous Asphalt Mix Design

Design characteristics	Results
Optimum bitumen, %	6.5
Air void, %	18.6
Cantabro abrasion loss, %	24.8
Coefficient of permeability, m/day	128.5
Indirect tensile strength, kPa	204.8

Effect of Crumb Rubber on Performance Characteristics

The effects of crumb rubber size and content on performance characteristics of porous asphalt mixtures were analyzed and discussed in this section.

Hydraulic Conductivity

Figure 6 illustrates the variations of permeability coefficient with rubber content for all three sizes (#4~#20, #20~#200, and #4~#200). Increasing the content of the crumb rubber significantly reduced the coefficient of permeability. Analyzing the crumb rubber in terms of particle size, it was indicated that #4~#20 rubber size had the maximum value of permeability coefficient. The addition of crumb rubber remained below limit value of permeability coefficient (100 m/day) except in the case of the 10% content of the #20~#200 crumb rubber. For #4~#200 crumb rubber, the addition of 20% crumb rubber similarly remained below limit value of 100 m/day. In all other cases, coefficient of permeability was provided over the limit value. These results show that the small size of rubber particle reduced air voids and the coefficient of permeability due to its surface area.

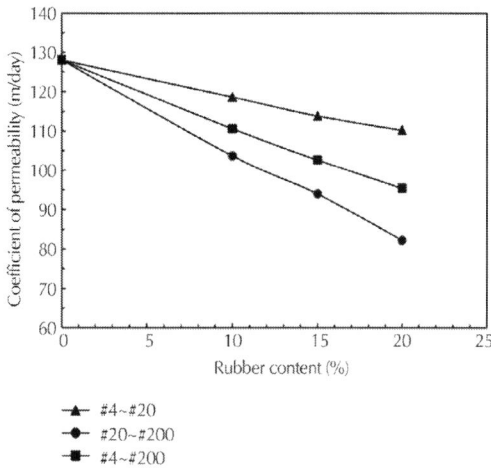

Figure 6: Effect of crumb rubber size and content on coefficient of permeability.

Cantabro Abrasion Loss

The experimental results were analysed as given in Figure 7. It can be clearly seen that Cantabro abrasion loss was achieved less than control specimens in the case of all contents of the #20~#200 crumb rubber. The addition of 10%, 15%, and 20% rubber content of #20~#200 showed an improvement of 19%, 27%, and 33%, respectively, compared to control. Particle loss of mixtures increased as the content of rubber increased for all rubber particle sizes. The optimum rubber concentration of 10% was determined. The degree of particle loss in the mixtures with #4~#20 and #4~#200 was observed to increase in comparison with control specimens. This increase of 130% reached at

20% content of #4~#20 rubber sizes in which the maximum particle loss occurred. #20~#200 mesh size include finer particles in comparison to the other mesh size (Figure 3). Finer rubber particles formed homogeny and rigid matrix (bitumen-fine aggregate, crumb rubber) due to the fact that they were uniformly mixed in the asphalt mixture. Because the coarse rubber particles in #4~#20 and #4~#200 mesh sizes caused increasing discontinuity in the bitumen film, it weakened the strength of the bitumen film. In addition, the #20~#200 mesh size increased the surface area which required surrounding particles by bitumen. Thus, decreasing bitumen film thicknesses tended to improve tensile strength of porous asphalt mixtures. As a result, Cantabro abrasion loss values decrease by using #20~#200 crumb rubber.

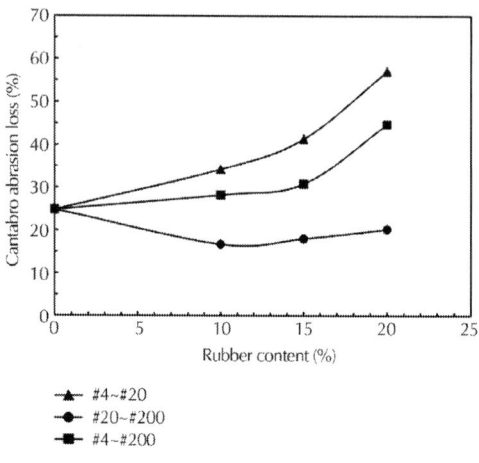

Figure 7: Effect of crumb rubber size and content on Cantabro abrasion loss.

Indirect Tensile Strength

The indirect tensile (IDT) strength test is very useful in deciding the performance of porous asphalt mixtures which depend on the cohesion of bitumen film. The variations of indirect tensile strength with rubber content for all three sizes are given in Figure 8. IDT strength test results were in compliance with Cantabro abrasion test results. The addition of crumb rubber significantly reduced the IDT strength except in the case of the 10% content of the #20~#200 crumb rubber. The case was the only one combination which increased the indirect tensile strength by 12% compared to the control specimens. IDT strength values were less than the control specimens for all rubber contents of #4~#20 and #4~#200. The maximum reduction in IDT strength by 32% was obtained in the case of the 20% content of the #4~#20 crumb rubber. In general, increasing the crumb rubber size has decreased the IDT strength of the porous asphalt mixtures. These results can be explained by the fact that the strong matrix was formed by homogeneously mixing small rubber particles in the mixtures and maintained an adhesion between the bitumen and the aggregate. It can be concluded that the cohesion and IDT strength of the bitumen film were negatively affected by the form of rubber particle as shown in Figure 4. Although larger crumb rubber particle shaped like pine needle rubber improves indirect tensile strength of dense graded asphalt pavement [24], it affects the performance of porous asphalt mixtures negatively because the porous asphalt mixtures have less fine aggregate.

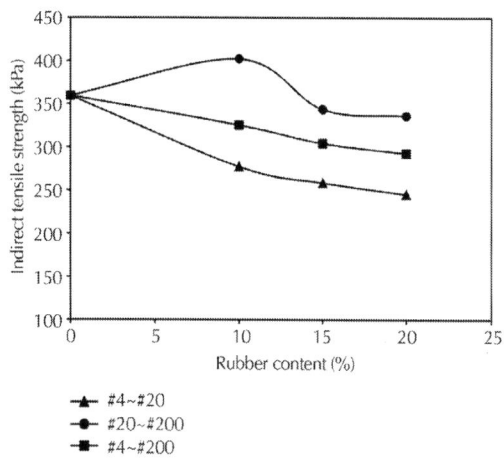

Figure 8: Effect of crumb rubber size and content on indirect tensile (IDT) strength.

Moisture Susceptibility

The moisture damage in the asphalt concrete pavements is an important problem. The porous asphalt pavement is particularly subjected to moisture damage more than dense graded asphalt mixtures. The moisture damage counts on the loss of adhesion between bitumen and aggregate surface or cohesion of the bitumen. The modification of bitumen or mixture is the most commonly used method in the improvement of the resistance to the moisture damage.

The TSR value for the control specimens was determined to be 58%. This value remained below the limit value of 70% because unmodified bitumen was used to prepare the control specimens. The unmodified bitumen was preferred to be employed for a better understanding of the effect of crumb rubber on the porous asphalt mixtures. An examination of the graphical presentations in Figure 9; the addition of crumb rubber significantly reduced the TSR value except in the case of the 10% content of the #20~#200 crumb rubber. For #20~#200 crumb rubber, the addition of %10 crumb rubber has obtained a similar TSR value with control specimens. TSR values were less than the control specimens for all rubber contents of #4~#20 and #4~#200. Increasing the crumb rubber size has substantially decreased the TSR values. The maximum reduction in TSR by 63% was obtained from the addition of the 20% content for the #4~#20 crumb rubber. As explained above, while the mixtures with small rubber particles improved performance of porous asphalt, the big size rubber particle negatively affected cohesion of bitumen and IDT strength. This effect increased largely the IDT strength of conditioned specimens. Therefore, the TSR values are significantly reduced compared to the control specimens.

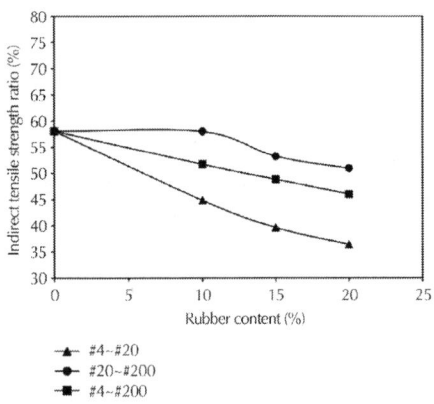

Figure 9: Effect of crumb rubber size and content on moisture susceptibility.

Resilient Modulus

The resilient modulus measured in the indirect tensile mode represents the elastic properties of asphalt mixtures under repeated load effectively. The resilient modulus was one of the most commonly used methods for measuring the stiffness modulus of hot-mix asphalt. The variations of resilient modulus with rubber content and rubber size were given in Figures 10 and 11, respectively. The addition of crumb rubber reduced the resilient modulus except in the case of the 10% and 15% contents of the #20~#200 crumb rubber. The maximum resilient modulus value was obtained in rubber content of 10% at #20~#200. In this case resilient modulus was increased by 22% compared to the control specimens. Increasing the crumb rubber size reduced the resilient modulus of the porous asphalt mixtures substantially. The maximum decrease in resilient modulus of 56% was determined in rubber content of 20% for #4~#200 crumb rubber. Resilient modulus values were decreased as crumb rubber content in the mixtures for all rubber particle sizes. The optimum content of crumb rubber was determined in 10%. This test results coincided with IDT strength and Cantabro abrasion test results. Therefore, the impact of crumb rubber size and content on resilient modulus can be interpreted as noted above.

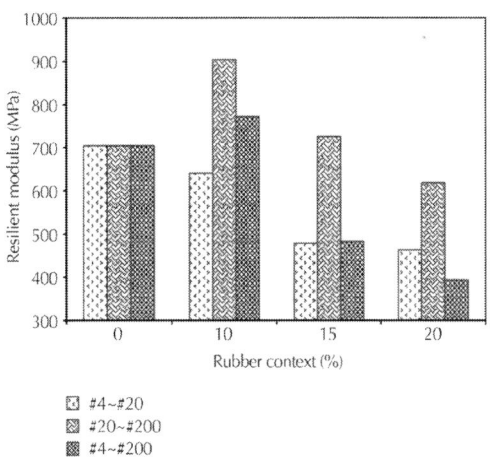

Figure 10: Effect of crumb rubber content on resilient modulus.

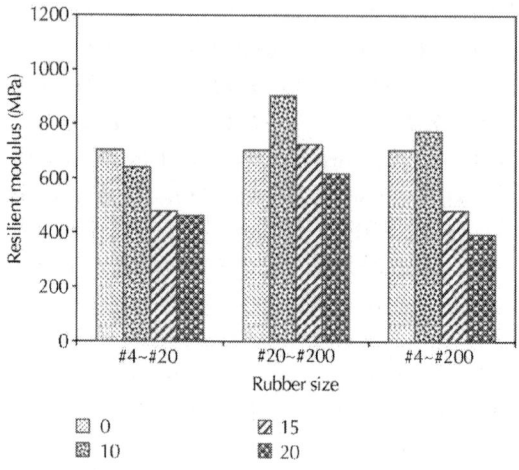

Figure 11: Effect of crumb rubber size on resilient modulus.

CONCLUSION AND RECOMMENDATION

This study presented an evaluation of recycled scrap tire obtained from tire recapping or tire buffing in the porous asphalt mixtures. The impacts of crumb rubber size and concentration on the performance characteristics of porous asphalt mixtures were investigated in this experimental study. Porous asphalt mixtures were modified by dry process using crumb rubber in three different grain size distributions (#4~#20, #20~#200, and #4~#200) and rubber contents of 10%, 15%, and 20% as weight of bitumen. The permeability, Cantabro abrasion, indirect tensile, moisture susceptibility (Modified Lottman), and resilient modulus tests were carried out in this experimental program. Based on the analysis of the results obtained from this study, the following conclusion and recommendation can be drawn.(i) The optimum bitumen content in the design of porous asphalt was determined to be 6.5%. Although the mix design of porous asphalt mixture has achieved design requirements with penetration grade bitumen of 50/70, the low values in Cantabro abrasion loss and IDT strength tests have demonstrated the necessity of modified bitumen.

(ii)Increasing the crumb rubber size and content significantly reduced the coefficient of permeability. The #20~#200 crumb rubber exhibited better performance than other grain size distributions, and the coefficient of permeability at the 10% rubber content remained over the limit value (100 m/day). The permeability test result showed that the small size of rubber particle reduced air voids and the coefficient of permeability due to its surface area.(iii)Cantabro abrasion loss of mixtures increased as the content of rubber increased for all rubber particle sizes. The optimum rubber concentration was determined to be 10%. All rubber contents of #20~#200 indicated an improvement of Cantabro abrasion loss compared with other rubber gradation. The maximum particle loss of 130% occurred at 20% content of #4~#20. The larger rubber particle size leads to an increase in the Cantabro abrasion loss on the account of the discontinuity in the bitumen matrix. (iv)While IDT strength of 12% only improved at 10% content of the #20~#200 crumb rubber, IDT values were significantly reduced for all rubber contents compared to the control specimens. In general, increasing the crumb rubber size decreased the IDT strength of the porous asphalt mixtures. The discontinuity in the matrix formed by large size rubber particle has a negative effect on the cohesion and IDT strength.(v)The addition of crumb rubber significantly reduced the TSR value except in the case of the 10% content of the #20~#200 crumb rubber. This value was similar to the TSR of control specimens of 58% which remained below limit value of TSR (70%) because of the fact that the control specimens were prepared with unmodified bitumen. Increasing the crumb rubber size has substantially decreased the TSR values. Modified Lottman Test results indicated that the addition of crumb rubber did not improve the moisture susceptibility performance.(vi)Resilient modulus test results corresponded to the IDT strength and Cantabro abrasion test results. Resilient modulus values were decreased with addition of crumb rubber to the mixtures for all rubber particle sizes. In addition, increasing the crumb rubber size decreased it substantially. The maximum resilient modulus value was obtained in rubber content of 10% at #20~#200. The optimum content of crumb rubber was determined in 10%.(vii)All test results show that #20~#200 mesh size which includes more spherically formed rubber particles than the other mesh sizes (Figure 4(d)) mixed uniformly in the mixtures and constituted a more rigid bituminous matrix. Therefore, it has partially improved the performance characteristics of porous asphalt mixtures. In contrast, larger rubber particle size leads to the discontinuity in the bitumen film. This larger size affected tensile strength of bitumen film negatively resulting in loss of performance. This effect largely increased the IDT strength of conditioned specimens due to loss of adhesion.(viii)The larger fiber-like shaped crumb rubber particles improve the indirect tensile strength of dense graded asphalt pavement, but it negatively affects the performance of porous asphalt mixtures depending on less finer aggregate in the porous asphalt mixtures.It can be recommended that the case of 10% content of the #20~#200 crumb rubber should be used to improve the performance characteristics of porous asphalt mixtures. The rubber content higher than %10 and the size larger than #20~#200 decrease the performance of the mixture significantly. Although the rubber particles size and concentration which was determined in this study by using dry process crumb rubber modification has provided the performance improvement for porous asphalt mixtures, the future investigation can be carried out about fractions of #20~#200 rubber gradation and concentrations less than 10%. Moreover, the wet process which requires additional cost in application should be studied to improve the performance of porous asphalt mixture.

REFERENCES

1. Rubber Manufacturers Association, "Scrap tire markets in the United States," 9th Biennial Report, Washington, DC, USA, 2009.

2. Federal Highway Administration, "User guidelines for waste and by product materials in pavement construction," Tech. Rep. FHWA-RD-97-148, 1997.

3. A. A. A. Molenaar, A. J. J. Meerkerk, M. Miradi, and T. van der Steen, "Performance of porous asphalt concrete," in Proceedings of the Annual Meeting and Technical Session (CD-ROM), Association of Asphalt Paving Technologists, Lino Lakes, Minn, USA, 2006.

4. A. Ruiz, R. Alberola, F. Perez, and B. Sanchez, "Porous asphalt mixtures in Spain," Transportation Research Record, vol. 1265, pp. 87–94, 1990.

5. P. M. Nelson and P. G. Abbott, "Acoustical performance of previous Macadam surfaces for high-speed roads," Transportation Research Record, no. 1265, pp. 25–33, 1990.

6. Q. Lu and J. Harvey, "Laboratory evaluation of open-graded asphalt mixes with small aggregates and various binders and additives," Transportation Research Record, no. 2209, pp. 61–69, 2011.

7. M. N. Partl, E. Pasquini, F. Canestrari, and A. Virgili, "Analysis of water and thermal sensitivity of open graded asphalt rubber mixtures," Construction and Building Materials, vol. 24, no. 3, pp. 283–291, 2010.

8. California Department of Transportation, Open Graded Friction Course Usage Guide, Sacramento, Calif, USA, 2006.

9. W. D. Cao, Z. Y. Yao, Q. S. Shang, Y. Y. Li, and Y. S. Yang, "Performance evaluation of large stone porous asphalt-rubber mixture," Advanced Materials Research, vol. 150-151, pp. 1184–1190, 2011.

10. T. Hsu, S. Chen, and K. Hung, "Performance evaluation of asphalt rubber in porous asphalt-concrete mixtures," Journal of Materials in Civil Engineering, vol. 23, no. 3, pp. 342–349, 2011.

11. R. C. Haas, E. Thompson, F. Meyer, and G. R. Tessier, "The role of additives in asphalt paving technology," Proceeding of the Association of Asphalt Paving Technologists, vol. 52, pp. 324–344, 1983.

12. J. W. Oliver, "Optimising the improvements obtained by the digestion of comminuted scrap rubbers in paving asphalt," Proceeding of the Association of Asphalt Paving Technologists, vol. 51, pp. 169–188, 1982.

13. B. Choubane, G. A. Sholar, J. A. Musselman, and G. C. Page, "Ten-year performance evaluation of asphalt-rubber surface mixes," Transportation Research Record, no. 1681, pp. 10–18, 1999.

14. R. G. Hicks and J. A. Epps, Quality Control for Asphalt Rubber Binders and Mixes, Rubber Pavements Association, Tempe, Ariz, USA, 2000.

15. National Asphalt Pavement Association, Porous Asphalt Pavement, National Asphalt Pavement Association, Lanham, MD, USA, 2003.

16. M. Backstrom and A. Bergstrom, "Draining function of porous asphalt during snowmelt and temporary freezing," Canadian Journal of Civil Engineering, vol. 27, no. 3, pp. 594–598, 2000.

17. AASHTO T275-91, Standard Method of Test for Bulk Specific Gravity of Compacted Bituminous Mixtures Using Paraffin-Coated Specimens, American Association of State and Highway Transportation Officials, Washington, DC, USA, 2005.

18. R. B. Mallick, P. S. Kandhal, L. A. Cooley Jr., and D. E. Watson, "Design, construction, and performance of new generation open-graded friction courses," NCAT Report 2000-01, National Center for Asphalt Technology, Auburn University, 2000.

19. Florida Department of Transportation, "Florida method of test fo measurement of water permeability of compacted asphalt pacing mixtures," Designation FM 5-565, 2012.

20. "ASTM D7064-08, standard practice for open-graded friction course (OGFC) mix design," in Annual Book of ASTM Standards, Road and Paving Materials, vol. 04.03, ASTM International, West Conshohocken, Pa, USA, 2008.

21. "ASTM D6931-07, standard test method for indirect tensile (IDT) strength of bituminous mixtures," in Annual Book of ASTM Standards, Road and Paving Materials, vol. 04.03, ASTM International, West Conshohocken, Pa, USA, 2007.

22. "Standard method of test for resistance of compacted asphalt mixtures to moisture-induced damage,"AASHTO T283-03, American Association of State and Highway Transportation Officials, Washington, DC, USA, 2007.

23. "ASTM D4123-82, standard test method for indirect tension test for resilient modulus of bituminous mixtures," in Annual Book of ASTM Standards, Road and Paving Materials, vol. 04.03, ASTM International, West Conshohocken, Pa, USA, 1995.

24. M. Tuncan, A. Tuncan, and A. Cetin, "The use of waste materials in asphalt concrete mixtures," Waste Management and Research, vol. 21, no. 2, pp. 83–92, 2003.

10

Landscape Engineering, Protecting Soil, and Runoff Storm Water

Mehmet Cetin

College of Engineering, Department of Civil and Environmental Engineering, Temple University, Philadelphia, Pennsylvania, USA

INTRODUCTION

Landscape engineering thinks about the application of mathematics and science to the creation of convenient outdoor living areas. These outdoor living areas are a consequence of the design and the construction process made feasible by landscape architects through landscape contractors. Beside landscape engineer's interested in containing site grading and drainage, earthwork calculations, and watersheds [1,12,13].

Landscape engineers employ engineering knowledge when designing and building spaces. They demand to know how to interpret contour maps which shows elevations and surface configuration by means of contour line sand and also consider how to interpret 2-D images, compute angles and grading requirements for pavements, parking lots, bridges, roads and other structures. Beside they understand the amount of fill needed for specific areas and figure out how water runoff and flow should affect their designs.

Developing and improving for landscape engineering [1,3,13,14]

- *Stormwater management*: building up bioswales, landscape materials used to gather and install runoff, rain gardens and porous asphalt.

- *Mitigation of the urban heat island effect*: diminishing the quantity of paved surfaces, building up green roofs and green walls, and using building materials with low reflectivity.

- *Wildlife habitat*: protecting habitat and building up green roofs and rain gardens.

- *Social spaces*: areas for walking, biking, gathering and eating.

- *Transportation*: growth expanding pedestrian accessibility, limiting vehicle speeds and encouraging the use of public transportation.

All above are using for the mixing plastic with asphalt on the pavement for the above reasons. One of them we are using porous plastic asphalt on the pavement for diminishing the sounds of road, and protecting the soil, and making runoff out the structure.

Porous plastic pavement which is a permeable pavement surface permits the action of storm water in order to infiltrate directly into the soil. It usually constructs with an underlying stone supply that temporarily stocks surface runoff before infiltrating into the subsoil. It takes the place of traditional pavement. There are various types of porous surfaces, including porous asphalt, permeable concrete and even grass or permeable pavers.

Porous plastic asphalt pavements suggest an alternative technology for stormwater management. It is varied from classical asphalt pavement designs so that the structure enables fluids to come frankly through it, decreasing or monitoring the amount of run-off from the surrounding area. By permitting precipitation and run-off to running out the structure, this pavement kind functions as an additional stormwater management technique. The mainly benefits of porous plastic asphalt pavements should contain both environmental and safety benefits including improved stormwater management, improved skid resistance, decreasing of spray to drivers and pedestrians, as well as a potential for noise decreasing. Porous plastic asphalt is applicable to many uses, including parking lots, driveways, sidewalks, bike paths, playgrounds and recreational courts. In addition, with proper maintenance, including regular vacuuming or pressure washing of the pavement surface to prevent clogging by sediments, porous asphalt can have a minimum service life of twenty years [15,16,17,18].

Bitumen is a valuable binder for road construction. Different grades of bitumen which is 30/40, 60/70 and 80/100 are available on the basis of their penetration values. The steady growth in high traffic intensity in terms of commercial vehicles, and the important variation in daily and weather temperature demand developed road characteristics.

Nowadays the waste plastics are to exist tremendous, as the plastic elements have been piece of daily life for using. They either blended with Municipal Solid Waste and/or dispose of landfill. If not recycled, their present disposal is either by land filling that the method has certain impact on the environment. Because of that an alternate use for the waste plastics is also the needed.

Thinner polythene carry bags are most adequately disposed of wastes, which do not draw attention the attending rag pickers to accumulate for forward recycling, for lesser value. These polythene bags are simply suitable with bitumen at specified conditions. The waste polymer bitumen mix may be prepared and a study of the properties can throw more light on their use for road laying.

Permeable pavements have been sufficient in behalf of environmental problems and corroborating sustainable green construction procedures. They are planned to support surfaces for parking lots and pedestrian roads that will permit some of the precipitation to filter into the ground, decreasing the bulk of stormwater runoff and revitalizing groundwater [15,16,17]. Permeable porous asphalt and concrete are the most ordinarily used elements in permeable pavements. Permeable asphalt uses practically the similar elements as conventional asphalt, order than the percent range of plastic added, and the size of the aggregate is used to stay thin, permitting for minimum particle packing [16,17].

Permeable asphalt and concrete has been successfully implemented for road, retaining walls, streets, driveways, sidewalks, parking lots, and slope protection for the past 30 years in many countries, including the United States [17]. None of any researchers, to the authors' knowledge, have inquired for mixing permeable plastic asphalt so far that implement plastic waste as a binder bonder for permeable pavements. This study presents an alternative sustainable technology, familiar as permeable plastic asphalt, which should be implemented for permeable pavements. Permeable plastic asphalt material is produced from plastic waste, porous aggregates, and asphalt. The research evaluated the mechanical and hydraulic characteristics of permeable plastic asphalt.

An asphaltic paving material involves 4%, 5% and 6% percent, of granular recycled plastic (LDPE with 1%, 3%, and 6%), which supplements the porous aggregate component (aggregate size of 3/8, 4, 8, 16) of the mixture. The material produces a structurally superior paving material and longer lived roadbed. The plastic can contain any and all residual classes of recyclable plastic. The material produces roadbeds of higher strength with less total asphalt thickness and having greater water permeability, and is most useful for all layers below the surface layer. A process of shredding or mechanically granulating preferably forms the paving product.

A porous plastic asphalt pavement has benefit for using road users and road side environments such as decreasing road noise, developing drainage function, and driving security situation. Furthermore, the worth of porous asphalt is to support skid resistance, particularly in the wet weather, that is distinctly better than that of dense traditional asphalt [19].

Conventional asphalt was not usually acceptable in the high temperature, humid, and traffic. As a result that some of issues were occurred like the pores were clogged, strength of drainage, rutting and scattering came off by traffic loading [20]. Large percentage of air voids of the mixture is adopted in order to maintain the drainage function. Nevertheless, in the way that the proportions of air void grow, strength feature of the mixture diminish. This is the main theorem of asphalt mixture. For solving this, landscape engineers think about the plastic for design and planning as binder.

The aim of this research is to assess strength, which means the performances of porous asphalt mixtures, for testing Resilient Modulus to find Indirect Tensile Strength (ITS). Furthermore, it measures the durability of porous asphalt mixture using Marshall Immersion test.

It is the aim of this chapter (book) to outline the elementary essential parts fundamentals associate with the design, plan, and construct of pavements and to mix with plastic techniques that will allow a Landscape Engineer to plan and design a pavement to suit a variety of situations. After experiments next books chapters will be considered to help environment and nature same as protect soil and recycling. Next study will keep going to research the better experiments about permeable surfaces for using different plastic materials. As a result it gains economic, environmental, and practical design and plan for Landscape Engineering.

Landscape Engineering

Landscape Engineering defined that is the art of developing land for people use and entertainment in such a manner as to obtain the utmost utility including with the utmost of beauty. It is a mistake to take into consideration the subject as implicated mainly with planting trees or to visualize its main function to be the supply of some decorative camouflage for some unsightly utility. Conversely, it has to be comprehended as a most important fundamental to life, and all art, which no utility would be necessary camouflage, and that every kind of artistic proceedings, instead of being without deep, has to be natural structural. Landscape engineering verifies the fundamental utilities, not as a necessary bad, but as necessarily well. In lieu of these utilities existing in the way of the objects, which the landscape engineer is succeed in doing, they turn out a most sufficient part of her/his own initiative. Apparently, once this point of view is admitted the Landscape Engineer and the others are definitely suiting for the same thing. Landscape Engineering impresses the main ideas, concepts, and techniques that cope with the functional, visual, and ecological perspectives of grading and landform cultivation. Landscape Engineering points out introduction to the processes, principles, and techniques of site engineering [1,3,13,21,22,23].

Students who pursue to study their career in Landscape Engineering programs learn about Landscape Design and Planning, Transportation Plan, Structural Landscape Design, Urban Planning, Site Planning, Ecological Design, Environmental Design, and Horticulture. Essential aims of all of subjects in Landscape Engineering pay attention to cover sustainability, and eco-friendly landscaping processes. They usually not only study with their specific area of studies like sustainable urbanism, environmental hazard management, historic preservation, ecological design but also focus in land development along with construction management, or a compound of land development and architecture [2,14].

Landscape engineer is that construction is the first step after designing and planning. Landscape engineer have to be continually consider to reach suitable adjustments between the operating cost, the construction cost, and the maintenance cost. For only during the construction period can the required savings be influenced at a minimum of expense. Landscape engineer think about completed without mention of subgrade conditions or specifications on material and method of placing while designing and planning pavement, parking lot, etc. Drains are often paid attention as releases for storm water during the catch basins although the drain is laid to accumulate the soil. With new experiment in planning is the topographical survey, usually worked out with care and precision during considers drain and protects soil.

Landscape engineer has to keep in mind when planning the pavement that it is a business proposition as well as a picture designing and planning. That the Landscape planned with an eye to operation and maintenance cost must in time, have the better financial situation to protect appearance. Meanwhile design and plans are begun; we have to never forget that budget is a factor in maintenance, whether the project is considered on the constant care principle. The complete design of the pavement is that while every element of pavement construction must be considered from the point of view of beauty and aesthetic value, the weight of construction, operation and maintenance must be found and recognized at the time of beginning [3,22].

A Landscape Engineer conducts and takes advantage of the strengths of nature for the benefit and satisfaction of man. For instance, if you were designing and planning a new landscape for your property and wanted to direct the winds to preserve your amusing areas, a

Landscape Engineer would help you make a determination which trees to get and where to plant them; thus, the wind could avoid your patio or courtyard. Landscape engineering involves with the application of mathematics and science to the constitution of practical outdoor living areas. These areas are a conclusion of the design and the construction process made possible by landscape architects, along with landscape contractors. Other main contemplates of landscape engineer's include drainage and site grading, watersheds and earthwork calculations. It is the Landscape engineer's act to vigorously interest in the design of the landscape, hence, to significantly supervise the creation of the landscape design. Landscape engineering exemplifies the traditional engineering elements of planning, operation, management, design and construction, and assessment. Before planning and designing, Landscape engineer should consider three main areas that focus on [4,5,24]:

- Landscape plan and design that Landscape Engineers interest planning the individual landforms so that they go along with the objectives arrange in the termination design phase. It is necessity that the Landscape engineer confers with the rest of the architectural team for carrying out a sufficient visual result once the construction process is finished.

- Termination design and plan that it implies setting targets as well as supporting the Landscape design of the project at hand. Landscape engineers who works very closely with the contractors and sub-contractors does for working with Landscape designers as well as the owner of a home or property for deciding what the desired look is considered together and the essential steps to be taken in order to success it.

- Functioning assessment that the reason of performance evolution of the Landscape engineer results in estimating liability and financial assurance of the landscape design and plan project. Thus, performance evaluation is vital to both the closure planning progress and performance evaluation.

Soil Protection

As we preserve air and water, which we use for breathing and drink, the soil is very important as well. The quality of soil takes care of ecosystems. Soil quality, which would be linked to water quality, is a measure of soil productivity. As a determination soil quality which particularly, the bulk of soil to operate within ecosystem borders to endure keep up environmental quality, biological productivity, and support plant and animal health [25,26,27].

Soil structure and water infiltration is very important for soil quality as well as aim attention at characteristics like organic matter content overall soil biological activity, nutrient availability, and total organic matter levels [25,27]. Soil can be protected from erosion, landslide, and compaction by mixing plastic materials. Soil conveys very easy underground by water. Soil, which is ancient, rock broken into sand, silt and clay is the recycled and continuously transforming.

The one of ways for protect and sustainably improve of soils is mixed cover pavement with plastic materials for prevent erosions of soil. Soil tests should be recommended for mixing with compaction plastic. It is evaluated compliance with hydrologic characteristics like drainage from plastic compaction.

Stormwater Runoff

Stormwater runoff comes off while precipitations from rain or snowmelt infiltrate the ground. Impermeable surfaces like pavement, driveways, parking lots, sidewalks, and streets restrain stormwater runoff from naturally permeate into the ground.

For understanding the impact of stormwater runoff firstly need to consider how important water cycles through the urban environment. The bulks of precipitation infiltrate forests flows slowly underground, are infiltrated in by natural progress, and eventually arrive at lakes, streams while is being its natural. When we design and plan to compress the land with roads, parking lots, and buildings, the natural progress of water infiltrate in the earth is diminished. The existing grasslands and forests are put in place of concrete, roofs, and asphalt that do not permit rain to infiltrate the earth. Rather than, the precipitations soak through as faster as directly into streams, storm drains, and all without the profit of filtration. Designing, planning and constructing with mixing plastic material help to infiltrate water on the driveway, parking lots, pavement in place of concrete or asphalt. Thus, they permit stormwater to quickly drain into the ground [28,29,30].

Stormwater is water that directly results from a rainfall event is not absorbed into soil and rapidly flows downstream, increasing the level of waterways. The flow of water results from precipitation and that occurs immediately following rainfall or as a result of snowmelt. Stormwater is the portion of rainfall that does not infiltrate into the soil. Rainwater and snowmelt that runs off impermeable surfaces rather than infiltrate into the soil through a drainage system of underground pipes, stormwater carries nutrients, fine soils, plant debris, drippings from vehicles, and other substances from the drainage basin most of lakes, ponds, and wetlands are connected to the stormwater system. Water collected in a system of pipes which drain roads and industrial or trade premises Stormwater may contain contaminants present on drained surfaces.

Stormwater is concerning about two issues that are the volume and timing of runoff water and prospective polluting. Stormwater is needed to flood control and water supplies. Also stormwater lead to be water pollution for conveying the water. As a stromwater, watermanagement on the pavements should probably do urban environments self-sustaining in terms of water. Stormwater is pollution because impermeable surfaces like parking lots, roads, buildings, compacted soil do not enable rain to drain into the ground. More runoff is constituted than in the unprogressive condition. Thus, it should consume waterways like flooding after stormwater collection system is overwhelmed by the additional flow. So the water is out of watershed way through little drainage the soil, the storm event [31,32].

Stormwater is a problem so that it could collect chemicals waste, mud, dirt, and other pollutants and infiltrate in the storm sewer system or directly to a river, lake, coastal water, stream, or wetland. Anything, which inserts a storm sewer system, is released untilled into the waterbodies we run for fishing, swimming, and supplying drinking water.

The results of pollution of stormwater runoff could have many unfavorable effects on animals, plants, people, and fish. Sediment that could blur the water and make it complicated or unimaginable for aquatic plants to grown could demolish aquatic habitats. The pollution of stormwater usually influences drinking water sources, which could transform to imitate human health and grow drinking water treatment costs.

The solution of pollution of stormwater that is permeable pavement consists of mixing plastic. Traditional concrete and asphalt don't tolerate water to infiltrate into the ground. Rather than these surfaces depend on storm drains to switch unwanted water, Permeable pavement systems permit rain and snowmelt to saturate with, diminishing stormwater runoff [28,32]. Stormwater runoff is without filtered water that arrives at oceans, streams, and lakes by means of streaming on impermeable surfaces that contain driveways, parking lots, roads, and roofs.

Correlation between Impermeable Pavement and Permeable Runoff

Impermeable pavement in a watershed occurs in growth permeable runoff. The barely 10 percent impermeable pavement in a watershed would occur in stream decline.

Stormwater is pollution so that impermeable surfaces like parking lots, compacted soil, roads, buildings, do not permit rain to flows off from the land in the streams, further runoff is caused to be than in the immature condition [28,31].

This further runoff would spoil streams and rivers as well as bring about flooding after the stormwater system is overflow by the extra flow so that the water is reveled out of the watershed through the storm event, barely drains the soil, fills groundwater, or stocks stream baseflow in dry weather. Contaminant inserting surface waters during rain that is lead to contaminate runoff. Daily people activities appear in sediment of pollutants on parking lots, farm fields, driveways, roads, lawns, roofs. As soon as precipitation begins, water flow off and eventually makes its way to a lake, ocean, and river.

A traditional city block creates more than five times more runoff than the forest because of impermeable pavement. The waste of penetration from city may come out with depth groundwater changes [28,31,32].

The present drainage systems, which accumulate runoff from impermeable surfaces like roads, parking lots, roofs insure that water is effectively transported to ways of water during pipes. As a matter affects little storms water occur in growth ways of water flows. Stomwater lead to some of issue that shows below.

- Impermeable Pavement
- Roads, sidewalks, rooftops, overly compacted soils
- Do not allow for natural infiltration of stormwater
- Increase temperatures (Heat Island Effect)
- Degradation of water quality and natural habitats
- Flooding, erosion and may reduce groundwater levels

Class of Plastic for Using Contemporarily

The simplest way for a user to label the class of plastic used in a product is to recognize the resin identification code, which is familiar with the material container code as well, is generally plotted, shaped or symbolized in or close to the middle of the bottom of the product. In accordance with the society of the plastics industry (SPI) resin identification code (or material code) is to systematize mutual plastic resins and their characteristics.

The identification of plastics made known in 1988 by The Society of the Plastics Industry Trade Association (SPI) that provides to build the companies for easy recycles make collect the consumer plastics during the common pathways for coming together recyclable stuffs from household waste. It is based on willing for plastic manufacturers, however, it has evolved into comparatively standard on plastic products sold in the U.S. and internationally. For example, the identification of plastics is in service and is affirmed by the Canadian Plastics Industry Association (CPIA) in Canada. It supports specifics on the identification by mean of its Environment, health and safety strategic unit and its Environment and Plastics Industry Council (EPIC) [7,33,34].

The aim of the identification provides to make it simpler for recycling to plastics. Furthermore, it determines consumers with an easy, handy method for identifying the class of plastic resin used to create a specific product. In conformity with SPI identifications, the number is intentionally located in an unnoticeable place on the product so that the company purpose is not to affect the consumer's buying determination, barely to assist the progress of recycling of the product.

According to SPI, there are seven different classes of plastics. Showing Figure 1, the identification numbers imprint on the bottom of plastic products that is a number inside of triangle represents to mean their identifications for recycling. Have you ever been curious about what the numbers inside the little recycling symbol mean on all of the plastic packaging and plastic products which we consume for using?

According to SPI the identification in 1988 reciprocates to the consider revising of the plenty recyclers side to side the countries. Here each class of plastics number and definition [7].

- PETE, PET (Polyethylene Terephthalate) is clarity, strength or toughness, barrier to gas and moisture, resistance to heat. It uses for consuming plastic soft drink and water bottles, beer bottles, mouthwash bottles, peanut butter and salad dressing containers, oven able film, oven able pre-prepared food trays.

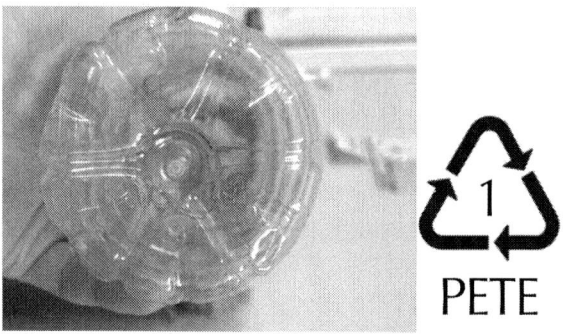

Figure 1: Samples of identification on the bottom of plastic water bottles.

- HDPE (High Density Polyethylene) is stiffness, strength or toughness, resistance to chemicals and moisture, permeability to gas, ease of processing, and ease of forming. It uses to make plenty classes of bottles. The bottles are clear, have good limit qualities and stiffness, and are quite appropriated to packaging products with a short shelf life like milk so that HDPE has good chemical resistance; it is used for packaging many household and industrial chemicals such as detergents and bleach. It uses milk, water, juice, cosmetic, shampoo, dish and laundry detergent bottles; trash and retail bags, yogurt and margarine tubs, cereal box liners.
- V, PVC (Vinyl, Polyvinyl Chloride) is versatility, ease of blending, strength or toughness, resistance to grease or oil, resistance to

chemicals, clarity. It has well chemical resistance, weather ability, flow typical features and constant electrical qualities. Products made from Vinyl can be both flexible and rigid. It uses toys, clear food and non-food packaging, shampoo bottles, medical tubing, wire and cable insulation, film and sheet; construction products such as pipes, fittings, siding, flooring, carpet backing, window frames.

- LDPE (Low-Density Polyethylene) is ease of processing, barrier to moisture, strength or toughness, flexibility, ease of sealing. It is used efficaciously in film uses because of its flexibility toughness, and approximate transparency, making it familiar for use in uses that heat sealing is essential. Furthermore, LDPE uses to procedure some flexible lids and bottles as well as in wire and cable uses. It uses squeezable bottles (honey, mustard), coatings for paper milk cartons and hot and cold beverage cups, container lids, toys, dry cleaning, bread, and frozen food bags.

- PP (Polypropylene) is strength or toughness, resistance to chemicals, resistance to heat, barrier to moisture, versatility, and resistance to grease or oil. It has good chemical resistance, is strong, and has a high melting point making it well for hot-fill liquids. This resin is brought to light in rigid and flexible packaging, fibers, and large pattern parts for automotive and consumer products. It uses containers for yogurt, margarine, takeout meals and deli foods, medicine bottles, bottle caps and bottles for ketchup. Furthermore, for packaging, its plenty of uses are in fibers, appliances and consumer products, containing strong applications like automotive and carpeting.

- PS (Polystyrene) is versatility, insulation, and clarity, easily foamed as known "styrofoam". It is clear, hard and brittle. Also, it has an approximately low melting point. General uses contain protective packaging, food packaging, bottles, and food containers. It is usually connected with rubber to make high impact polystyrene (HIPS) that is used for packaging and constant uses necessity stiffness. It uses compact disc cases, food- service applications, grocery store meat trays, egg cartons, aspirin bottles, cups, plates, and cutlery.

- Other is dependent on resin or combination of resins. Use of this number represents which a package is made with a plastic other than the six listed above, or is made of more than one plastic and used in a multi-layer combination. It uses usually shows the exits of polycarbonate which a hard, clear plastic used to make baby bottles, water pitchers, nalgene brand water bottles, three and five-gallon reusable water bottles, food containers, some citrus juice and ketchup bottles, compact discs, cell phones, automobile parts, computers, three and five-gallon reusable water bottles, some citrus juice and catsup bottles, oven-baking bags, barrier layers and custom packaging.

DEVELOPMENT OF DESIGN

Design Factors

The area of pavement design is vigorous in which ideas are steadily changing as new data evolve into achievable. For mixing plastic pavement that are many majority of design and plan applicable, since alternatives relating to sustainability, suitability of designs and plans alter from area to area. Especially, supplies that are applicable for construction and foundation of pavements have a higher impress on design and plan. There are, nevertheless, fundamentals of design that are mutual to all problems irrespective of other uncontrollable situation [35,36,37,38,39].

The plan and design of pavement embrace with a work of soils and paving materials, their action under load, and the plan and design of pavement to convey which load under all hydraulic and weather situations. All pavements obtain their eventual support from the underlying subgrade. As a result, a comprehension of elementary plastic materials, pavement design and soil mechanics is necessary. Landscape Engineers are familiar that efficiency of pavement that are connected to a large volume upon the types of plastic, soils over that the pavement is designed and constructed, therefore, in relationships pavement efficiency between subgrade types are built. On the whole, the experiments of mixing plastic pavement demonstrated that pavement designed and constructed over plastic displayed higher degrees of distress than those designed and implemented over traditional pavement. Frost process and unfavorable drainage situations were observed early as two of the primary reasons of pavement lapse.

However, many landscape engineers are made use of standard cross components for most pavements. It means that a road, even though it crossed several mixing plastic and soil types, was designed and implemented using a constant thickness. The foreseeing was usually confirmed on the rest of economics. Beginning of 1980s, people recognized that the traditional pavement affects the environment and nature. All of thriving technological elements such as cars, bike for becoming a simple life causes to influence the pavements lapse [40,41,42,43].

Conventional Pavements

AASHTO have been in charge for various tests, which roads designed and implemented in United States as well as some of state highway departments have implemented test pavements for the aim of assessment the influence of load and elements on pavement design.

There is a small suspicion that the outcomes of test results have had extreme effect on current design ideas. Furthermore, efficiency of example pavements in service has had important effect on design. This is not shocking, if one thinks that it is hard to do if not unreasonable to judge entirely design ideas in the laboratory. In addition, it has been familiar with for quite a while that reader belief in the final analysis imposes the competency in some of the design.

Definition Pavement Types

Basically, with regard to history of pavements have been classified two main types which are flexible and rigid. Flexible pavements comprise asphalt. Nowadays flexible pavement is very important with mixing plastic so that makes permeable areas.

Plan, design, and construction of permeable pavements have altered rather importantly in the last decades. On account of the current traditional plan, design, and construction pavements come up severe higher traffic levels, wheel loads, pavement lapse. A growth use of balanced is base and subbase. Balancers such as asphalt, plastic are repeatedly used to grow the structural strength of the pavement by growth rigorously. Because of the reason an extremely concentrated effort was made in the last several years to develop a more fundamentally based design analysis for asphalt.

Permeable Plastic Asphalt

Some of researchers assessed the impact of moisture susceptibility on porous asphalt samples [44]. Samples were based on wet and dry conditions and then tested for indirect tensile strength test (ITS). Results showed that ITS decreased noticeably when the sample was immersed in water.

Purpose of Plastic Asphalt

The aims of the study are need to do those. First of all; Disposal of waste plastic is a major problem, non-biodegradable, burning of these waste plastic bags causes environmental pollution. Secondly need is it mainly consists of low-density polyethylene, and to find its utility in bituminous mixes for road construction. Thirdly, Laboratory performance studies were conducted on bituminous mixes. Laboratory studies proved that waste plastic enhances the property of the mix, and improvement in properties of bituminous mix provides the solution for disposal in a useful way.

Waste plastics like polythene carry bags, etc. on heating usually at approximately 160°C. According to thermo gravimetric results has demonstrated that gas evolution isn't found during the temperature rank of 130 to 180°C. Furthermore the mellowed plastics have a binding feature. Therefore, the melted plastics can be implemented as a binder that they should be blended with binder like bitumen to improve their binding feature. As a result it should be a good modifier for the bitumen, implemented for road construction.

Function of the Mix Plastic Asphalt

The growth of plastic city wastes has affected to wide and creative technologies that incorporate recycled plastics in miscellaneous uses. Scientists and Departments of Transportation have been interested in different researches regarding to the feasibility, economic and ecological impact and the complete efficiency of recycled plastic in connected landscape engineering projects. For instance, recycled thermoplastics like PET, HDPE, and LDPE, have been implemented in porous asphalt mixtures to put in place of aggregates with specified diameters [45,46,47,48,49,50]. Conclusion from these researches demonstrated improvement in strength, durability, and fatigue life. Nevertheless, the scale of improvement is a capacity of the plastic types and amount. The rest of researchers implemented recycled plastic strips to mechanically stabilize and aggregates by mixing plastic shreds with aggregates to compaction to defeat inadequacies in grading and diminishing the plasticity index [51,52,53]. The scale of enhancement was affected by many factors like the class and volume of shreds, and aggregate classes. Furthermore, creative and innovative study has cause to the development of new mixed elements using recycled plastic waste for miscellaneous uses. The mixed is produced by heating and blending the absorption elements, recycled plastic, which flakes, shared, or unprocessed, granulates and by products to a highlighted temperature. The heated combination is then compacted into a particular mold to found a final product. The features of the mixed rely on the pressure, class of recycled plastic and granulates [54,55,56].

MATERIAL PROPERTIES

Recycled Low-Density Polyethylene (LDPE)

The supplier for Recycled Low Density Polyethylene (LDPE) provided the test properties of the material with respect to density, tensile strength at break, elongation at break, impact strength, and melting point of the material as shown below Table 1 [8,10]. Recycled low-density polyethylene (LDPE), which is identification number four, was gathered together and implemented in this research. The cleaned LDPE has shredded as shown below Figure 2.

Recycled Low-density polyethylene (LDPE) material is used extensively to produce tote bags for domestic goods. These bags become solid waste after their use for short periods and cause serious waste disposal problems. To solve this environmental problem, and at the same time to improve the drain down and other related engineering properties of the porous asphalt mixture, reclaimed from LDPE bags was used in this investigation as additive in porous asphalt mixtures. LDPE material in shredded of used is as added ingredient.

Table 1: Recycled Low Density Polyethylene Features (LDPE)

Recycled Low Density Polyethylene Features (LDPE)	
Mechanical Properties	
Yield Strength	15-20 MPa
Elongation @ break	600-650 %
Bending Strength	10-40 MPa
Young's modulus (E)	200-400 MPa
Shear modulus	100-350 MPa
Tensile Strength (σ_t)	8-12 MPa
Physical Properties	
Density	910-928 kg/^{m}3
Thermal expansion	150-200 e-6/K
Water absorption	0.005-0.015 %
Melting Point	248 °F 120 °C
Thermal conductivity	0.3-0.335 W/m.K
Melting temperature	125-136 °C
Maximum Temperature	176 °F 80 °C
Minimum Temperature	58 °F 50 °C
Specific heat (c)	1800-3400 J/kg.K

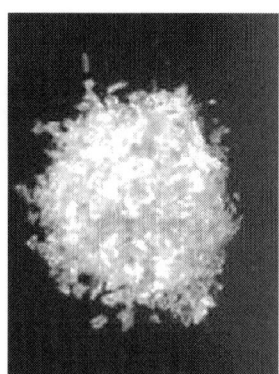

Figure 2: Shredded for recycled Low Density Polyethylene (LDPE).

Porous Aggregate

Crushed limestone was chosen as the course aggregate for mixing LDPE. Bulk samples were sieved in conformity with the sieve sizes for AASHTO No. 8. According to Figure 3 demonstrates the gradation for aggregates that Porous aggregates confirming to the sizes 3/8 in., Nos. 4, 8, 16 (AASHTO No 8) were used for mixing with permeable plastic asphalt. Aggregates maintained on each sieve were washed, dried for 24 hours in 110°C in the oven and then located into their respective batches by sieve maintained. This procedure provided regenerate samples to meet AASHTO No. 8. Furthermore, it made to be better control over the gradation of each sample, so that gradation has important impact on the engineering and physical features of an aggregate mixed.

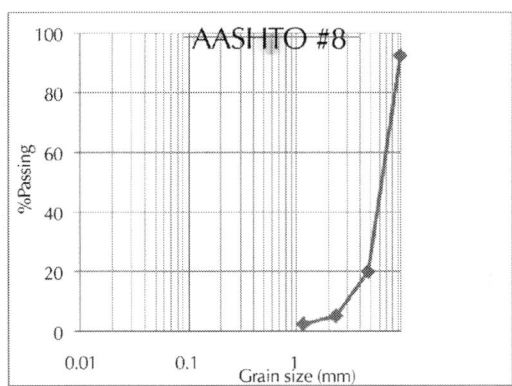

Figure 3: Grain size distribution of AASHTO No. 8.

Bitumen

"PG 68-22" was used in the porous asphalt mixture [9]. Bitumen was mixed with Low-density polyethylene (LDPE) and porous aggregates. Mixes were prepared for three (4%, 5%, and 6% bitumen) percentages of bitumen. Each % of bitumen has 1%, 3%, and 6% LDPE. Obviously, four types of mixtures with three different percent of bitumen of 4%, 5% and 6% were used at a mixing and compacting temperature of

160°C. These are: Without LDPE, 1% LDPE, with 3% LDPE, and with 6% LDPE.

SAMPLES PREPARATION, COMPACTION, AND TESTS

Mixing of Shredded Waste Plastic (LDPE), Aggregate AND Bitumen

The aggregate mix is heated to 160°C in oven, and similarly the bitumen is to be heated up to a maximum of 160°C. Plastic waste is shredding for mixing bitumen and aggregate to coat the plastics effectively. After that, put them in oven 160°C in order to mix and compact simultaneously. For protecting the moisture the spacemen, it compacted immediately after took out oven with approximately 160°C.

Mix Design by Marshall Method Marshall Test

Use of the processed plastic bags is as an additive in bituminous concrete mixes. The processed plastic was used as an additive with heated bitumen in different proportions (ranging from 4 to 6 % by weight of bitumen) and mixed well by hand, to obtain the modified bitumen. The properties of the modified bitumen were compared with ordinary bitumen.

Varying percentages of waste plastic by weight of bitumen was added into the heated aggregates. Marshall sample with varying waste plastic content was tested for stability. Maximum value of stability was considered as criteria for optimum waste plastic content. The optimum modified binder content fulfilling the Marshall Mix design criteria was found to be 4, 5, and 6 % by weight of the mix, consisting of 1,3, and 6 % by weight of processed plastic added to the bitumen. In order to evaluate the ability of the mix prepared with the bitumen to withstand adverse soaking condition under water, Marshall Stability tests were conducted after soaking in water at 60°C for 24 hours.

Porous Plastic Asphalt for Preparation and Compaction

Shedder LDPE was simultaneously composed with binder and aggregates for heating and mixing approximately 160°C for two hours so that a uniform was achieved. Mixed plastic porous asphalt was poured and compacted into a mold, which is 4 inches diameter and 2.5 inches height, using a steel shovel. The Marshall test procedure was used for designing porous mix by compacting the sample with 50 blows on one face by Marshall hammer, at varying binder contents. The mixture design trials used asphalt content in the range of 4 – 6 %, by total weight of the mixture, excluding the weight of the fibers, with 1% increments. The LDPE fibers were added to the porous mixtures at a dosage rate of 1,3, and 6 % based on total mixture weight. The compacted samples were extracted from the mold when they had sufficiently cooled. After compaction, samples were be kept in the hot (conditioned) and cold (unconditioned) waters. Resilient Modulus and permameter test conducted. Samples were tested for hydraulic conductivity and indirect tensile strength. Figure 4 demonstrates the preparation of the Porous plastic asphalt samples.

Laboratory Tests

Porous plastic asphalt samples were tested for hydraulic conductivity and indirect tensile strength. Hydraulic conductivity tests were operated using a falling head approach in conformity with a proceeding particularized in the researchers [57,58]. The indirect tensile tests were conducted in accordance with the ASTM C 6931-07 test methods.

Sample Preparation

Plastic Waste (Low Density Polythylene-LDPE), Bitumen PG-22

Thermoplastic

Melting temperature

Bitumen mixed Plastic Waste (LDPE)

Porous Aggregates

Water goes through the plastic asphalt

Spaces after compaction

Figure 4: Sample preparation for permeable plastic asphalt that mixing of LDPE, porous aggregates, binder to create.

From equation 1 calculate hydraulic conductivity for testing in the falling head permameater test. After used Permameters to measure k, use this following formula. Calculate the hydraulic conductivity of the sediment by using the following formula:

$$k = \frac{V \times L}{(h_0 - h)A \times t} \times \ln\frac{h_0}{h}$$

(1)

where K = hydraulic conductivity of the sediment sample [L]/[T]

V = volume of water that passed through the sample [L]^3

L = sample length [L]

h_0 = height of top mark above outflow port [L]

h = height of bottom mark above outflow port [L]

A = cross sectional area of sample. For the NEIU permameters, this is 31.65 cm^2. [L]^2

t = total time for discharge [T]

The indirect tensile test that is one class of tensile strength test implemented in order to stabilize elements. The test has been run on asphalt-stabilized elements [35,59,60,61]. The test has many advantages, the most obvious being simplicity of test procedure. From equation 2 calculate ITS for testing in the resilient modulus (MR) test. After used with the diametrical Mr test (repetitive indirect tensile modulus test) to calculate ITS, as using the tensile strength St of the material is given by:

$$S_t = \frac{2P\max}{\pi td}$$

(2)

Where P= total applied load (lb)

t= sample thickness (in)

d= sample diameter (in)

TEST RESULTS

Permeability

According to Table 2 demonstrates the hydraulic conductivity (k) of the porous plastic asphalt samples. From Table 2, it has provided that the results of k diminished with the growth of porous plastic mixing. For instance, when the results of k mixed %4 binders with %1 LDPE were 0.204 in/s, the result of mixing %3 LDPE was 0.193 in/s. Obviously, while mixing with % of LDPE was increasing, the result of k was diminishing. Figure 5 represents the results of permeability of porous plastic asphalt with mixing %1 LDPE.

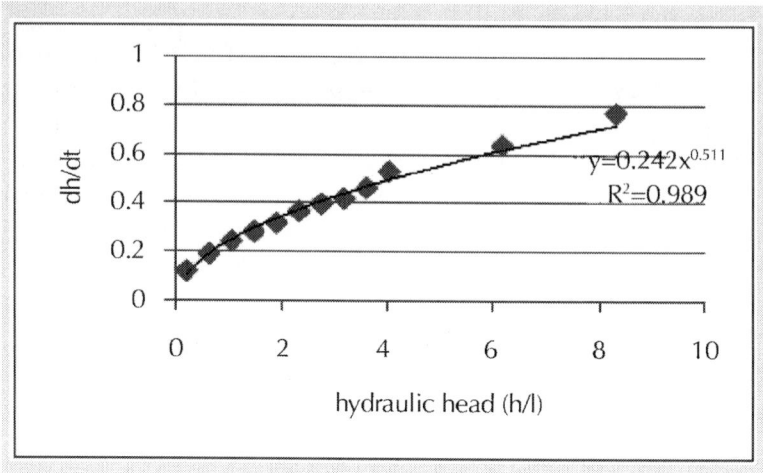

Figure 5: The results of k with mixing %1 LDPE.

Table 2: The results of hydraulic conductivity

Binder	LDPE	k (in/s)
% 4	%0	0.217
	%1	0.204
	%3	0.193
	%6	0.178
% 5	%0	0.259
	%1	0.245
	%3	0.234
	%6	0.223
% 6	%0	0.301
	%1	0.294
	%3	0.288
	%6	0.272

Indirect Tensile Strength (ITS)

Use this test method to determine the tensile strength of compacted bituminous mixtures. The porous plastic mixed tested as a conditioned and unconditioned. After compaction, samples were kept in the hot water (conditioned situation) and cold water (unconditioned situation). The objective of this test was to measure the water resistance of the mixture after immersion for 24 hours at 60°C. After that, testing the resilient modulus provided the results of ITS. The purpose of results was to evaluate the resistance of porous plastic asphalt mix on plastic deformation. Furthermore, from equation 3 Tensile strength ratio (TSR) calculated that divided by conditioned to unconditioned situation. TSR < 70% considered Susceptible to Moisture. As a result it provides that the performance of strength of porous plastic asphalt. Moisture Susceptibility of Porous Plastic tested in accordance with ASTM C 6931-07. Samples were cured at room temperature, 100°C and 160°C for 24 hours. Samples dimension is Diameter=4", Height=2.5".

$$TSR = \frac{S1}{S2} \tag{3}$$

Where:

S1 = conditioned set (wet)

S2 = unconditioned set (dry)

According to Table 3 demonstrates that the results of unconditioned decreased with the growth of porous plastic mixing. For instance, when the results of mixed %5 binders with %3 LDPE were 57 psi, the result of mixing %6 LDPE was 54 psi. Obviously, while mixing with % of LDPE was increasing, the results of conditioned ITS was decreasing. On the contrary, the result of conditioned increased simultaneously with the growth of porous plastic mixing. Figure 6 represents that the result of conditioned ITS diminished while unconditioned ITS increased. According to the results of TSR is considered to susceptible to moisture. It provides that %3 LDPE and over is very strength.

Table 3: The results of Indirect Tensile Strength test (ITS)

Binder	LDPE	Unconditioned (psi)	Conditioned (psi)	TSR (con/ uncon)
% 4	%0	67	45	0.67
	%1	58	52	0.90
	%3	46	56	1.22
	%6	41	63	1.54
% 5	%0	71	54	0.76
	%1	62	59	0.95
	%3	57	62	1.09
	%6	54	68	1.26
% 6	%0	73	63	0.86
	%1	69	67	0.97
	%3	66	71	1.08
	%6	61	75	1.23

The results of k and ITS of porous plastic asphalt mixtures were within the approximately predictable ranked come up in the literature for conventional asphalt showing Table 4. It is an indication that porous plastic asphalt should be used as a sustainable alternative for permeable pavements.

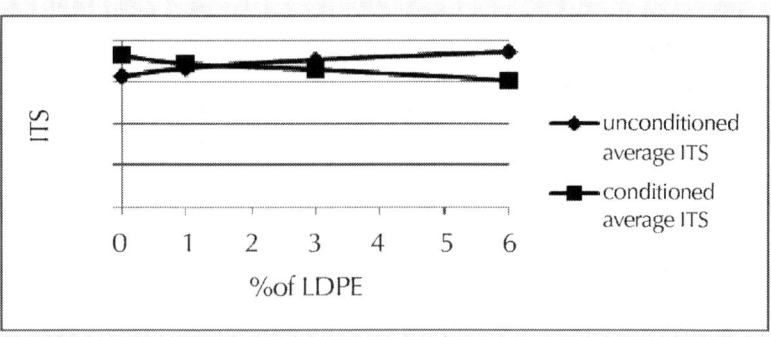

Figure 6: The results of ITS with mixing %6 LDPE.

Table 4: Summary of ITS porous asphalt from literature

References	ITS ranges (psi)
[62]	29.29-69.76
[63]	18.17-38.82
[64]	253.9-377.09
[65]	29-65.26
[66]	26.1-130.53
[19]	18.30-55.90

DISCUSSIONS AND CONCLUSIONS

Permeable Plastic Asphalt samples with different percentage of plastic to aggregate ratio were composed and then tested for hydraulic conductivity (k) and indirect tensile strength (ITS). Results represented which k and ITS results of samples were within the predictable results observed in the literature for porous pavements. In addition to it was come up with that the k, ITS values diminished as the percentage of plastic waste increased. The research results of Permeable Plastic Asphalt demonstrated that Permeable Plastic Asphalt should be a sufficient recycle and stormwater runoff with important reasonable economic and ecological associations.

This research was taken in charge to evaluate the hydraulic and mechanical features of innovative porous plastic asphalt for implementing in permeable pavements. The test represented that the result of hydraulic conductivity diminished with the growth of porous plastic mixing. On the other hand, while the results of unconditioned decreased with the growth of porous plastic mixing, the result of conditioned increased simultaneously with the growth of porous plastic mixing. Furthermore, the results of TSR are the best result over 1 for susceptible to Moisture. The results of experiments were approximately come across that expected from literature. According to results that demonstrate porous plastic asphalt could be implement sustainable alternative pavements. It also provides also recycling.

This research concentrated on the hydraulic and mechanical properties of a permeable pavement like permeable plastic asphalt.

Permeable Plastic Asphalt was composed of plastic waste, aggregates, and asphalt. Permeable Plastic Asphalt should provide a sufficient method for decreasing stormwater runoff, contributing a structural pavement sufficient for pedestrian and vehicular loadings. Furthermore, Permeable Plastic Asphalt should take the part of the currently ongoing recycling aims as a critical role that support to deflect a majority of plastic from landfills and incinerators.

The laboratory test indicates that aggregate, binder with mixing LDPE affected the results of k and ITS. Thus, permeability and strength of porous plastic asphalt is getting better with mixing LDPE. It proves that porous plastic asphalt help to diminish storm water runoff times diminish urban heat island effects. A new pavement increases for the sustainability of the nature that will be benefit users for many years. The design of plastic pavements contains developed pedestrian and public transportation as well as parking lot, driveway, bridges.

The results represents porous plastic asphalt should be implemented as a sustainable alternative for permeable pavements. Porous plastic asphalt is a peerless choice in that it undertakes two environmental problems that decreasing stormwater runoff and prevent to fill out with plastic waste at landfills. In the way that we keep up our way to green building and construction, porous plastic asphalt is new approach on the way to eco-friendly improvement. The innovative technology comes to grips with two environmental problems that are plastic waste and stormwater runoff. It provides to prevent a large quantity of plastic waste at landfills and incinerators, thus the plastic waste uses fro recycling. Also it decreases stormwater runoff and decreases the use of natural resources.

Permeable plastic pavement has whole with its permeability should be determined by valid void. Valid void should be directly implement to mixing ratio of permeable plastic pavement that both take control sufficiently the forming of run off and restrict urban flood.

All above results that Landscape Engineering considers that the progress of permeable plastic uses to efficiently integrate the mixing permeable plastic pavements with land use planning. Using recycling service with very powerful has approached to create pleasing environments in the world. According to test results that the advantages of plastic asphalt provide that is stronger road with increased plastic, better resistance during stormwater, without stripping and rutting, develop binder and better linking of the mix. Besides it support that is the strength of the road is increased. Using permeable plastic asphalt that is the cost of road construction diminish the maintenance cost of road gradually diminish as well. Obviously, as the plastic mix with pavement for using, the disposal of waste plastic will no longer be issue. As a result of that, using plastic helps to decrease in pores in aggregate save bitumen and help recycling.

Consecutive chapters of this research will count profoundly on the outcomes of the test pavements mentioned above, as well as efficiency data issued in the research. Detailed representing results of the several researches projects will be debated during the research.

FUTURE RESEARCH

Effectively managing the collection, separation and processing of plastic waste can limit the environmental damages limited by eliminating the waste from our streets. Thus, we can prevent to fill the landfill with plastic waste when we mix the plastic with the other elements with soil, asphalt and cement in order to use future studies.

Laboratory tests and real life implementation will study that the life expectancy of a plastic polymer road as compared to a conventional road. Future study will need to study for expanding of life expectancy for plastic. This study proved that investigates, summarizes preliminary results, and debates key properties to be considered for future alternative pavement. Future researchers will keep going to research the better permeable pavement for economic, environmental and nature.

A long-term monitoring project to document changes in performance, evaluation of different maintenance strategies, and lifecycle costs of permeable plastic asphalts is recommended for future research. Future experience is based on designs that provide to improve future properties such as increased new materials, and developed construction and maintenance activities.

ACKNOWLEDGEMENT

The author gratefully acknowledged Dr. Naji Khoury for his guidance and help.

REFERENCES

1.   G. T Mckenna, Sustainable Mine Reclamation and Landscape Engineering. PhD Thesis in Geotechnical Engineering, Department of Civil and Environmental Engineering, University of Alberta, Edmonton, 2002

2.   American Society of Landscape ArchitectureASLA. http://www.asla.org/ppn/Article.aspx?id=1206&terms=landscape%20engineeringaccessed 5 September 2012

3.   American Society of Landscape ArchitectureASLA. http://www.asla.org/uploadedFiles/CMS/Government_Affairs/Public_Policies/Licensure_Definition_of_Practice.pdfaccessed 5 September 2012

4.   E Zube, Landscape Planning Education in America: Retrospect and Prospect. Landscape and Urban Planning, 131986367378

5.   The LA GroupLandscape Architecture and Engineering P.C. http://www.thelagroup.comaccessed 8 September 2012

6.   A Valderrama, L Levine, S Yeh, and E Bloomgarden, Financing Stormwater Retrofits in Philadelphia and Beyond, Natural Resources Defense Council (NRDC), February 2012http://www.nrdc.org/water/files/StormwaterFinancing-report.pdfaccessed 11 September 2012).

7.   The plastics industry trade associationSPI Resin Identification Code- Guide to Correct Use. Washington. DC. http://www.plasticsindustry.org/AboutPlastics/content.cfm?ItemNumber=823&navItemNumber=1125accessed 2 November 2011

8.   LDPEHusky, 2008Polly America, 2000 West Marshall Drive, Grand Prairie, Texas 75051, phone 800 527 3322, press 7654, www.poly-america.comaccessed 2 July 2011).

9.   Asphalt 6422cas no 8052-42-4 PG-2010NuStar Asphalt Refining, LLC, 4Paradise Rd. Paulsboro, NJ 08066 phone 856 224 7405.

10. V. S Punith, and A Veeraragavan, Characterization of Ogfc Mixtures Containing Reclaimed Polyethylene Fibers. Journal of Materials in Civil Engineering, ASCE, 341, March 2011

11. A Khan, Gangadhar, Mohan M., and Raykar V., Effective Utilization of Waste Plastics in Asphalting of Roads. Project Report Prepared Under The Guidance of R. Suresh and H. Kumar, Dept. of Chemical Engg., R.V. College of Engineering, Bangalore, 1999

12. G Zaro, K C Nystrom, A Bar, A. U Alvarez, and A Miranda, Tierras Olvidadas: Chiribaya Landscape Engineering and Marginality in Southern Peru, The Society for American Archaeology, Latin American Antiquity 21(4), 3553742010

13. J Corner, Representation and Landscape: Drawing and Making in The Landscape Medium, Word & Image 8, 246, July-Sept. 1992

14. American Society of Landscape ArchitectureASLA Security Design Symposium Abstracts. Chicago, IL, July 2004http://www.asla. org/uploadedFiles/CMS/Resources/securitydesignabstractfinal. pdfaccessed 2 July 2011).

15. L. M Haselbach, S Valavala, and F Montes, Permeability Predictions for Sand-Clogged Portland Cement Pervious Concrete Pavement Systems, Journal of Environmental Management, 8120064249

16. B Huang, H Wu, X Shu, and E. G Burdette, Laboratory Evaluation of Permeability and Strength of Polymer-Modified Pervious Concrete, Construction and Building Materials, 24, 8188232010

17. P. D Tennis, M. L Leming, and D. J Akers, Pervious Concrete Pavements. Porland Cement Association. 2004http://myscmap. sc.gov/marine/NERR/pdf/PerviousConcrete_pavements. pdfaccessed 2 July 2011).

18. F Montes, S Valavaka, and L Haselbach, A New Test Method for Porosity Measurements of Portland Cement Pervious Concrete, Journal of ASTM International, 2(1), 2005

19. B Subagio, D Kosasih, Busnial, and Tenrilangi D., Development of Stiffness Modulus and Plastic Deformation Characteristics of Porous Asphalt Mixture Using Tafpack Super™, Proceedings of the Eastern Asia Society for Transportation Studies, 58038122005

20. H Nakanishi, T Kawanaka, L Ziqing, and H Baocun, Study on Improvement in Durability and Function of Porous asphalt Pavement, Road Construction, Japan, 2001

21. S Daniels, and H Repton, Landscape Gardening and the Geography of Georgian England, New Haven, CT: Yale University Press, 1999

22. D Harris, and D Hays, On the Use and Misuse of Historic Landscape Views, in Representing Landscape Architecture, ed. Marc Treib., New York: Taylor and Francis, 22412008

23. K Rieder, Modeling, Physical and Virtual, in Representing Landscape Architecture, ed. Marc Treib., New York: Taylor and Francis, 2008

24. F. A Waugh, Landscape Engineering in the National Forests. US. Department of Agriculture Forest Service. 1918http://archive.org/stream/landscapeenginee00waug#page/n1/mode/2upaccessed 5 October 2012).

25. J. W Doran, and T. B Parkin, Defining and Assessing Soil Quality. In: J.W. Doran et al. (eds.) Defining Soil Quality for a Sustainable Environment. Soil Science Society of America, Madison, WI, Special Publication 35, 3221994

26. B Mollison, Increases in Porosity Enhance Infiltration and Thus Reduce Adverse Effects of surface runoff, Permaculture: A Designer's Manual, Tagari Press, 1988

27. T Hubbard, Encyclopedia of Surface and Colloid Science 3Santa Barbara, California Science Project, Marcel Dekker, New York, 2004

28. K Beven, and E Robert, Horton's Perceptual Model of Infiltration Processes, Hydrological Processes, Wiley Intersciences DOIhyp 5740, 2004http://earth.boisestate.edu/jmcnamara/files/2011/10/KBeven_HP2004.pdfaccessed 23 August 2012).

29. Stormwater Pennsylvaniahttp://www.stormwaterpa.orgaccessed 25 August 2012

30. D Busco, and G Lindsey, An Annotated Bibliography of Stormwater Finance Resources, Center for Urban Policy and the Environment School of Public and Environmental Affairs Indiana University- Purdue University Indianapolis, April 29, 2002http://

stormwaterfinance.urbancenter.iupui.edu/PDFs/Biblio%20 4%2029%2002.pdfaccessed 26 August 2012).

31.   J. H Lee, and K W Bang, Characterization of Urban Stormwater runoff. Water Research. 346April 200017731780http://www.sciencedirect.com/science/article/pii/ S0043135499003255accessed 23 August 2012).

32.   M. E Dietza, and J. C Clausenb, Stormwater Runoff and Export Changes with Development in a Traditional and Low Impact Subdivision. Journal of Environmental Management, Microbial and Nutrient Contaminants of Fresh and Coastal Waters, 874June 2008, 560566http://www.sciencedirect.com/science/article/pii/ S030147970700103Xaccessed 23 August 2012

33.   Canadian plastics industry associationhttp://www.plastics.ca/ home/index.phpaccessed 2 November 2011

34.   Canadianplasticsindustryassociation Environmentalsustainability. http://www.plastics.ca/EnvironmentalSustainability/index.php/ accessed 2 November 2011

35.   E. J Yoder, and M. W Witczak, Principles of Pavement Design. Second Edition. John Wily And Sons, Inc, New York. 1975

36.   W. E. A Acum, and L Fox, Computation of Load Stresses in a Three-layer Elastic System, Geotechnique, 22933001951

37.   F Mccullough, A pavement Overlay Design System Considering Wheel Loads, Temperature Changes, and Performance, Graduate Report, ITTE, University of California, Berkeley, 1970

38.   F. N Hveem, Pavement Deflections and Fatigue Failures, Highway Research Board Bulletin 342, 1962

39.   E. S Barber, Application of Triaxial Compression Test Results to the Calculation of Flexible Pavement Thickness, Proceedings, Highway Research Board, 1946

40.   L Fox, Computation of Traffic Stresses in a Simple Road Structure, Department of Scientific and Industrial Research, Road Research Technical Paper 9, 1948

41.   McLeodand Norman W., An ultimate Strength Approach to Flexible Pavement Design, Proceedings, Association of Asphalt Paving Technologist, 1954

42. Hornerand Raymond C., Effect of Base Course Quality on Load Transmission Through Flexible Pavements, Proceedings, Highway Research Board, 1955

43. L. A Palmer, and E. S Barber, Soil Displacement under Circular Load Areas, Proceedings, Highway Research Board, 1940

44. L. D Poulikakis, and M. N Partl, Evaluation of Moisture Susceptibility of Porous Asphalt Concrete using Water Submersion Fatigue Tests, Construction and Building Materials, 2009

45. A Hassani, H Ganjidoust, and A Maghanaki, Use of Plastic Waste (Poly-ethylene Terephthalate) in Asphalt Concrete Mixture as Aggregate Replacement, Waste Manage Res 2005

46. S Hinislioglu, and E Agar, Use of Waste High Density Polyethylene as Bitumen Modifier in Asphalt Concrete Mix. Materials Letters, 582004267271

47. S. E Zoorob, and L. B Suparman, Laboratory Design and Investigation of Proportion of Bituminous Composite Containing Waste Recycled Plastics Aggregate Replacement (Plastiphalt), CIB Symposiumon Construction and Environment Theory into Practice, Sao Paulo, Brazil, November, 2000

48. S. E Zoorob, Laboratory Design and Performance of Improved Bituminous Composites Utilising Plastics Packaging Waste", Conference on Technology Watch and Innovation in Construction Industry, Belgium, Building Research Institute, Brussels, Belgium, April, 2000

49. A Conigliaro, and P Watson, Determining the Best Formulation for a Unique Asphalt Cold Patch Product Made with # 37Rigid Plastic Aggregate, Chelsea Center for Recycling and Economic Development Technical Research Program, January 2000

50. R. B Mallick, and M Teto, Evaluation of Use of Manufactured Waste Asphalt Shingles in Hot Mix Asphalt, 2000Chelsea Center for Recycling and Economic Development Technical Research Program, July 2000.

51. K Sobhan, and M Mashnad, Mechanical Stabilization of Cemented Soil-Fly Ash Mixtures with Recycled Plastic Strips, Journal of Environmental Engineering, ASCE, October 2003

52. K Sobhan, Stabilized Fiber-Reinforced Pavement Base Course With Recycled Aggregate, Ph.D. Dissertation, Northwestern University, Evanston, Illinois, 303 pages, June, 1997

53. J. K Cavey, R. J Krizek, K Sobhan, and W. H Baker, Waste Fibers in Cement-Stabilized Recycled Aggregate Base Course Material. Transportation Research Record 1486, Transportation Research Board, Washington D.C., 971061995

54. MeyersIII, Swartz, J, Nathaniel, Kurczewski, N., and Kurczewski, M., Recyclable Composite Materials Articles of Manufacture and Structures and Method of using Composite Materials, U.S. Patent 62006

55. R Malloy, and M Kashi, and C Swan, Fly Ash/Mixed Plastic Aggregate and Products Made Therefrom, U.S. Patent 62003

56. E Balkum, Aggregate Using Recycled Plastics, U.S. Patent 62002

57. T. F Fwa, S. A Tan, and C. T Chuai, Permeability Measurement of Base Materials using Falling-Head Test Apparatus, Transportation Research Record, 161594991998

58. N. N Khoury, C. N Khoury, and Y Abousleiman, Soil Fused with Recycled Plastic Bottles for Various Geo-Engineering Applications. ASCE Conf. Proc. 309, 42, 2008

59. M. M Frocht, Photoelasticit, 2John Wiley and Sons, New York, 1957

60. W. O Hadley, W. R Hudson, and T. W Kennedy, Evaluation and Prediction of The Tensile Properties of Asphalt-treated Materials, Highway Research Board Annual Meeting, Washington, D.C., 1971

61. J. W Kennedy, and W. R Hudson, Application of the Indirect Tensile Test to Stabilized Materials, Highway Research Board Annual Meeting, Washington, D.C., 1968

62. S. N Suresha, V George, and A. U. R Shankar, Effect of Aggregate Gradations on Properties of Porous Friction Course Mixes, Material and Structure, 2010

63. A Setyawan, Design and Properties of Hot Mixture Porous Asphalt For Semi-Flexible Pavement Applications, Media Technic Sipil, July 2005

64.   Q Liu, E Schlangen, A Garcia, and M. V. D Ven, Induction Heating of Electrically Conductive Porous Asphalt Concrete. Construction and Building Materials, 2010

65.   D Shen, C Wu, and J Du, Laboratory Investigation of Basic Oxygen Furnace Slag for Substitution of Aggregate in Porous Asphalt Mixture. Construction and Building Materials 23, 2009

66.   F Xiao, W Zhao, T Gandhi, and A. N Amiskhanian, Influence of Antistripping Additives on Moisture Susceptibility of Warm Mix Asphalt Mixture. Journal of Materials in Civil Engineering, October 2010

Chapter 11

Method of Quantification of Hydrated Lime in Asphalt Mixtures

V. Mouillet[a], D. Séjourné[a], V. Delmotte[a], H.-J. Ritter[b], and D. Lesueur[c]

[a]Laboratoire d'Aix-en-Provence, CEREMA/DTerMed, Pôle d'activités Les Milles, Avenue Albert Einstein, CS 70499, 13593 Aix-en-Provence Cedex 3, France

[b]Bundesverband der DeutschenKalkindustrie e.V. (BVK), Annastraße 67 – 71, 50968 Köln, Germany

[c]Lhoist R&D, 31, rue de l'Industrie, 1400 Nivelles, Belgium

ABSTRACT

Hydrated lime has been known as an additive for asphalt mixtures for a long time and is now considered as an additive that increases asphalt mixture durability. It has been extensively used in the past 40 years in the USA, and is being increasingly used in most European countries, in particular Austria, France, the Netherlands, the United Kingdom and Switzerland. Given this context, it is necessary to have a fast and reliable quantification method of the hydrated lime content in an asphalt mixture.

A German method was used in order to do so. The test method consisted first in recovering the filler from the asphalt mixture using the usual solvent extraction method (EN 12697-1). Then, 1 g of the recovered filler was titrated with a 0.5 M HCl solution using a method adapted from EN 459-2. The test method was validated on an AC 10 mixture manufactured in the laboratory. The nominal content was 2.0% hydrated lime based on the dry aggregate. The measured content was found to be 1.7%, in reasonable agreement with expected results.

As a result, the hydrated lime content in an asphalt mixture can be evaluated. An estimate of the precision of the method is also given thanks to an international round robin test, showing that the repeatability of the method is close to 0.7% and its reproducibility 4.5% in terms of $Ca(OH)_2$ content in the recovered filler.

INTRODUCTION

Hydrated lime has been known as an additive for asphalt mixtures from their very beginning [1], [2] and [3]. It experienced a strong interest during the 1970s in the USA, partly as a consequence of a general decrease in bitumen quality due to the petroleum crisis of 1973, when moisture damage and frost became some of the most pressing pavement failure modes of the time. Hydrated lime was observed to be the most effective additive [4] and as a consequence, it is now specified in many States and it is estimated that 40 Mt of asphalt mixtures are now produced in the USA each year with hydrated lime [5].

Given its extensive use in the past 40 years in the USA, hydrated lime has been seen to be more than a moisture damage additive [3], [6], [7], [8] and [9]. Hydrated lime is known to reduce chemical ageing of the bitumen [3], [6], [7] and [8]. Furthermore, it stiffens the mastic more than normal mineral filler [3], [6],[7] and [8], an effect only observed above room temperature [3], that impacts the mechanical properties of the asphalt mixture.

Given that all the above mixture properties impact the durability of asphalt mixtures, the use of hydrated lime has a strong influence on asphalt mixtures durability [10]:

- North American State agencies estimate that hydrated lime at 1–1.5% in the mixture increases the durability of asphalt mixtures by 2–10 years, that is by 20–50% [5],
- The French Northern motorway company, Sanef, currently specifies hydrated lime in the wearing courses of its network, because they observed that hydrated lime modified asphalt mixture have a 20–25% longer durability [11],
- Similar observations led the Netherlands to specify hydrated lime in porous asphalt [12], a type of mix that now covers 70% of the highways in the country.

As a result, hydrated lime is being increasingly used in asphalt mixtures in most European countries, in particular Austria, France, the Netherlands, the United Kingdom and Switzerland. The mechanisms behind this improvement in durability have recently been reviewed [13].

In this context, it is somewhat surprising that the problem of the quantification of hydrated lime in asphalt mixtures has only attracted interest in the recent years. As detailed below, only two methods can be found in the literature, both published less than 10 years ago: the first one comes from the USA and the second one, from Germany. In this work, we chose to use the German method in order to quantify hydrated lime in an asphalt mixture, because the testing equipment needed to put it into practice is usually already found in most road laboratories. In addition, its cost is very limited and the method could therefore be easily made available to a large number of control laboratories.

Therefore, this article first describes the published methods to quantify hydrated lime in an asphalt mixture. Then, it details the German method which has been first validated and then tested for repeatability and reproducibility in a new European round robin test.

BACKGROUND: AVAILABLE METHODS TO QUANTIFY HYDRATED LIME IN ASPHALT MIXTURES

US Method

The US method was developed by the Federal HighWay Administration (FHWA – [14] and [15]). It consists in measuring the Fourier Transform Infra-Red (FTIR) spectrum of the filler and quantifying the hydrated lime content from the peak intensity at 3,640 cm^{-1} corresponding to calcium hydroxide (Fig. 1). Calcium carbonate peaks at 1,390 cm^{-1} and can be unmistakably separated from hydrated lime (Fig. 1).

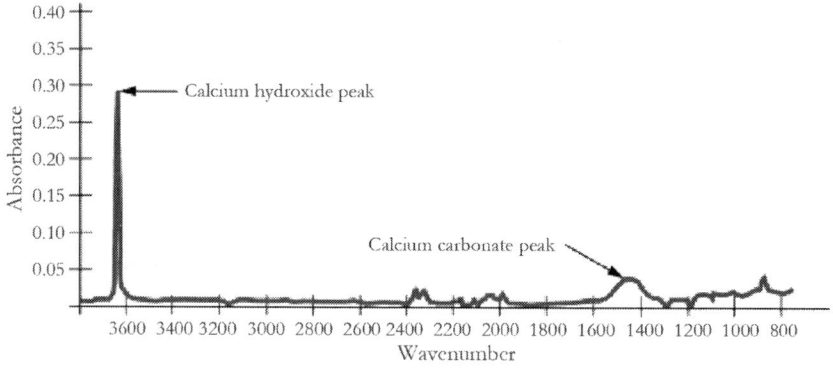

Figure 1: FTIR spectrum of hydrated lime (absorbance in arbitrary unit and wavenumber in cm^{-1} – adapted from [14]).

The analysis was shown to be easily performed by using 15–20 g of dust recovered by hammer drilling through an asphalt mixture with a 9.5 mm tungsten carbide bit [14] and [15].

Interestingly, measurements on 10 years old materials from Nevada showed that hydrated lime could still be detected after several years of traffic and weather exposure [14].

German Method

As explained in more details below, the German method [16] and [17] is very simple and derives from the lime characterization methods detailed in EN 459-2 [18]. In fact, the German method separates three different characterization sub-methods:

- Hydrated lime purity.
- Hydrated lime content in a mixed filler.
- Hydrated lime content in the filler recovered from an asphalt mixture.

The test consists in a hydrochloric acid titration of a suspension of the product to be tested. The acid has to be diluted (0.5 M) when mixed or recovered fillers are concerned, in order to adapt for a lower basicity. The filler is recovered from an asphalt mixture using solvent extraction of the bitumen as described in EN 12697-1 (usually using trichloroethylene or tetrachloroethylene as a solvent – [19]). The suspension to be titrated is then obtained by blending 1 g of recovered filler to 150 ml of water, 10 ml isobutanol and 5 ml of a surfactant solution (1 g sodium dodecylsulfate and 1 g polyethyleneglycol dodecylether in 100 ml water). The surfactant solution is needed only when recovered filler is tested, in order to wash out the filler from remaining bitumen or solvent from bitumen extraction. The colored indicator is phenolphthalein (0.5 g in 50 ml ethanol, completed to 100 ml by water). Titration rate is 12 ml/min initially, but decreases to 4 ml/min near the equivalence point. The method was shown to work with all types of fillers, including limestone filler [16].

A first national round robin test was performed in Germany with 12 laboratories [16]. The repeatability (in terms of absolute % weight of hydrated lime in the filler) was 0.52% and the reproducibility was 0.91% for a mean value of 27.3 wt.%.

The method was validated on samples taken out of cores 1.5 years after construction (Table 1 – [16]). The SMA 0/8 S mixes were made either with normal filler or with mixed filler containing 25% hydrated lime and the results are given in Table 1[16].

Table 1: Results of the validation of the German quantification method (after [16])

Section	Nominal hydrated lime content (wt.%)	Measured hydrated lime content in recovered filler (wt.%)
1	0	0.9
2	0	0.7
3	25	29.2
4	25	26.0

Note also that a study using different methods showed that the titration method was equivalent to the sugar method, which is the reference one in EN 459-2. Interestingly, the comparison based on asphalt mixtures made with different aggregates showed that part of the hydrated lime was not fully recovered, because of the hydrated lime – aggregate reactions (Fig. 2). As a result, these reactions were more important for basalt aggregate (about 60% recovery), than moraine (about 80%) and limestone filler (about 90%).

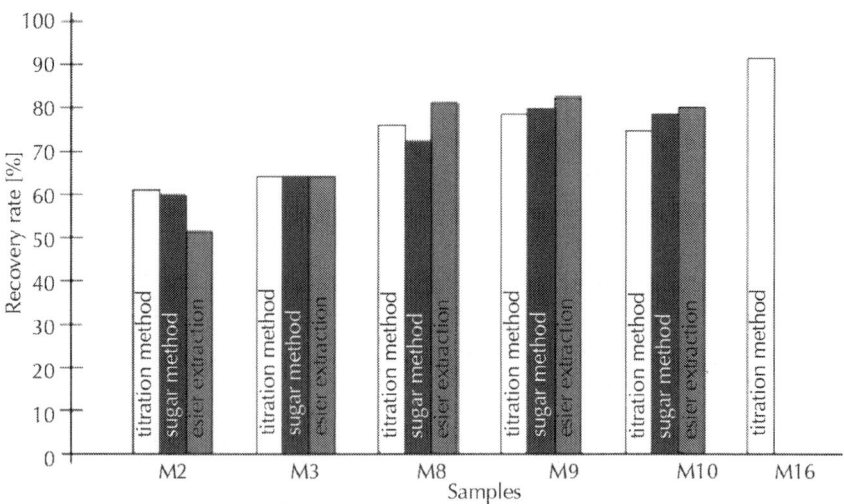

Figure 2: Percentage of hydrated lime eventually detected using three different chemical methods (adapted from [16]). "Titration method" refers to the direct titration following the German method [17] described at length in the text. "Sugar method" refers to the titration of a saccharose extract of the filler

to be tested and "Ester extraction", to an ethyl-acetoacetate extract. The materials were asphalt mixtures with different fillers mixed with hydrated lime: M2 and M3 with basalt filler (respectively 5% and 20% hydrated lime), M8 and M9 with moraine filler (respectively 5% and 20% hydrated lime), M10 with 67% moraine and 33% limestone filler (25% hydrated lime) and M16 with limestone filler (20% hydrated lime). The recovery rate is the ratio of measured hydrated lime over nominal hydrated lime content.

Other Methods

It is worth mentioning that Thermo-Gravimetric Analysis (TGA) could be also used in order to quantify $Ca(OH)_2$ in recovered filler. This is because the mass loss due to $Ca(OH)_2$ dehydration occurs in the 400 °C range [20] when common minerals found in asphalt mixtures do not normally undergo changes in this temperature range. However, there is a risk of confusion in the case of dolomitic aggregate, because the decarbonation of $MgCO_3$ occurs in the same temperature range [20]. Therefore, the method was not tested further but could potentially be used for equipped laboratories if they are confident that the aggregate is not dolomitic.

MATERIALS AND PROCEDURES

Materials

In order to validate this method of quantification of hydrated lime (accuracy of the titration and repeatability), the first stage of this study has consisted in the titration of known concentration solutions of:

- *Pure components*: two samples of CL90 S (according to EN459-1) hydrated lime from Dugny (France) and Flandersbach (Germany) have been tested. A limestone filler coming from the EJL La Nerthe quarry (France) was used as well.

- *Mixed fillers manufactured in laboratory*: preparation using a laboratory divider of mixed filler with known concentrations of hydrated lime: 25%, 50% and 75% of lime. These hydrated lime contents have been chosen in order to be representative of the mixed fillers available on the European market. For example, in

Germany or the Netherlands, mixed fillers containing 20% or 25% of hydrated lime are mostly used, while, in France, they go up to 75% of hydrated lime. In general, the mineral filler used with hydrated lime is calcium carbonate but it could be another mineral filler. For this study, only one limestone filler has been selected in order to mimic using a mixed filler prior to introduction into the asphalt mix. Also, calcium carbonate being one of the most acid soluble rock that is commonly found in road aggregate, it serves as a good reference in order to check the risk of dosing some of the filler as hydrated lime.

- Recovered filler from typical continuous asphalt mixture: preparation in laboratory of asphalt mix including filler with different known concentration of Dugny hydrated lime as a substitute for added filler followed by quantitative recovery of the recovered filler according to the European standard method EN 12697-1 (Asphalt Analysator method).

Initially, 3 asphalt mixtures were manufactured in the laboratory. They were based on a continuous semi-coarse asphalt concrete AC 10 35/50 according to EN 13108-1 (Table 2). The mixture was made of crushed river gravel from the Durance Granulats quarry (France). The 35/50 bitumen came from Total La Mède refinery (France). The reference formula was prepared in the laboratory using 3.8% of the limestone filler from EJL La Nerthe (Table 2). The other two formulas were based on mixed fillers with 50% and 100% Dugny hydrated lime respectively, in substitution for the added filler.

Table 2: AC 10 35/50 formula used for the filler extraction. The corresponding total filler content in the mixture is calculated

Fraction	Content in mixture (%)	Measured filler content in the fraction (%)	Actual filler content in the mixture (%)
0/2	22.7	15.0	3.4
2/6	34.1	0.7	0.2
6/10	34.1	0.3	0.1
Filler	3.8	94.0	3.6
35/50 bitumen	5.3	–	–
Total filler			7.3

German Method

The German test method [17] is used for the determination of the calcium hydroxide content in hydrated lime and mixed fillers for hot mix asphalt. It can be used also for fillers recovered from asphalt mixture.

The test method consists first in recovering the filler from the mix using the usual solvent extraction method (EN 12697-1). Then, two steps are performed to titrate the recovered filler:

- Disperse the filler sample in a mixture of water, isobutanol and tenside solution in order to clean the filler (to remove the bitumen and/or the extraction solvent maybe still present).
- Titrate in the alkaline range the calcium hydroxide content with hydrochloric acid using a method adapted from EN 459-2. But, it has to be noted that for mixed fillers and recovered fillers, it is necessary to determine the blank value of the used filler material.

This test method is very simple, easy to perform by all laboratories and unexperienced operators.

As 1 mol of $Ca(OH)_2$ reacts with 2 mol of HCl, the calcium hydroxide content expressed as $(Ca(OH)_2)$ in mass fraction in %, is given by the following equation:

$$\% \, Ca(OH)_2 = 100.F.37.05.(C_1.V_{eq})/(1000.m_1)$$

where: C_1 is the concentration of hydrochloric acid (mol/l) (note: as the concentration of solutions might deviate in time, a corrective factor has to be determined by titration with a base prepared by weighing), V_{eq} is the volume equivalent of hydrochloric acid (ml), m_1 is the mass of taken filler sample (g), F is the factor of the acid, being 1 for an acid at the right purity at ambient temperature.

International Round Robin Test

The round robin test was performed in the summer of 2011. 37 laboratories from all over Europe participated (Table 3), 20 being control or research laboratories of lime producers and 17 being road laboratories involved in the formulation and control of asphalt mixtures. Each laboratory was asked to quantify the hydrated lime content using the German method, in the same recovered filler with known content of hydrated lime.

Table 3: List of laboratories involved in the round robin

Country	Number of laboratories	Lime producer	Road laboratory
Austria	7	1	6
Belgium	6	3	3
France	1	0	1
Germany	13	11	2
Italy	4	1	3
Netherlands	1	0	1
Norway	2	1	1
Spain	3	3	0
Total	37	20	17

The round robin consisted in first manufacturing in the laboratory a new asphalt mixture containing a given hydrated lime content, namely 29.9%. This figure was obtained by adding 6% filler holding 70% hydrated lime with 94% purity, in a mixture having in total 12.4% filler (i.e. added filler plus filler from the aggregate fractions – Table 4). This nominal lime content was not disclosed to the laboratories. All the filler was recovered from the mixture following EN 12697-1 at Cerema Aix-en-Provence and was then sent to the Bundesverband der Deutschen Kalkindustrie e. V. (BVK) in Köln (Germany). There, the filler was prepared into 25-g specimen sent to each participating laboratory. Along with the filler, the needed tensides (sodium dodecylsulfate and polyethyleneglycol) were added. This was done so because the local availability of the surfactants was not insured in all countries.

Table 4: AC 10 surf 35/50 formula used for the filler extraction for the round robin. The corresponding total filler content in the mixture is calculated

Fraction	Content in mixture (%)	Measured filler content in the fraction (%)	Actual filler content in the mixture (%)
0/2	44.0	15.0	6.6
2/6.3	20.0	0.7	0.1
6.3/10	24.4	0.3	0.1
Filler	6	94.0	5.6
35/50 bitumen	5.6	–	–
Total filler			12.4

Each laboratory received the test method in three languages (English, French and German). It was asked that each laboratory would duplicate the quantification of calcium hydroxide in the received filler specimen, and report both results. Together with the final result expressed in calcium hydroxide content in the filler, we requested that the laboratory also report the mass of filler used for the trial and the volume of acid used. This way, potential deviations coming from the use of more concentrated/diluted acids could be easily detected.

RESULTS

Titration of Calcium Hydroxide Content of Suspensions with Known Concentration

Purity of Hydrated Lime

The specifications on hydrated lime (EN 459-1) oblige to state the purity that can affect the titration. Indeed, the hydrated lime can come from a low purity limestone which would yield a low $Ca(OH)_2$ content. Also some of the $Ca(OH)_2$ could have recarbonated due to extended storage in wet conditions, which would also decrease the available $Ca(OH)_2$

content. Consequently, the purity of components has been determined using the German method (Table 5). For each of the pure components, 5 repeatability trials have been performed to assess the standard deviation of the analysis. As expected, no calcium hydroxide content has been found in the limestone filler and the available $Ca(OH)_2$ (purity) of the 2 hydrated limes was around 95%. The standard deviation of the analysis was very good, around 0.3%.

Table 5: Calcium hydroxide content of the pure components (hydrated lime and limestone filler). sd: standard deviation

Pure components	Calcium hydroxide content
Limestone filler	0%
Dugny hydrated lime	94.7%
	(sd: 0.2%)
Flandersbach hydrated lime	94.5%
	(sd: 0.3%)

Titration of Mixed Filler with Hydrated Lime

As defined in Section 3.1, mixed fillers have been manufactured in the laboratory adding known concentrations of hydrated lime (25%, 50% and 75%) to limestone filler. The different mixed fillers have been prepared with both kinds of hydrated lime.

The accuracy of the titration of mixed fillers with Dugny hydrated lime was quite good as the relative deviation in relation to theoretical content was less than 2% for all lime contents (Table 6). Note that the standard deviation of the analysis (performed on 5 repeatability trials) was very good, around 0.6%.

Table 6: Calcium hydroxide content of mixed fillers with Dugny and Flandersbach hydrated limes. "Theoretical content" refers to the expected value, obtained by correcting the nominal content for the purity of hydrated (Table 5). sd: standard deviation

Samples	Nominal lime content (%)	Calcium hydroxide content		
		Measured content	Theoretical content (%)	Relative deviation from theoretical content (%)
Dugny hydrated lime	25	23.3% (sd: 0.6%)	23.7	1.7
	50	46.5% (sd: 0.5%)	47.4	1.9
	75	69.7% (sd: 0.6%)	71.0	1.8
Flandersbach hydrated lime	25	23.3% (sd: 0.6%)	23.7	1.7
	50	46.5% (sd: 0.5%)	47.4	1.9
	75	69.7% (sd: 0.6%)	71.0	1.8

For the mixed fillers with Flandersbach hydrated lime, the accuracy of the titration was quite similar to the previous one, although with a somewhat higher deviation (3.0%) for the 25% lime content. The standard deviation was slightly higher at 1.3%. The relative deviation in relation to theoretical content was comprised between 1% and 3% (Table 6).

Titration of Recovered Filler from Typical Continuous Asphalt Concrete

The fillers from the 3 variants of AC 10 35/50 asphalt mixture was recovered with the Asphalt Analyzator (EN 12697-1). About 8% filler was recovered, as expected from mixture composition (see Table 2). All recovered fillers have been quantitatively recovered according to

European standard method and titrated according to German method (Table 7).

Table 7: Calcium hydroxide content of recovered fillers from the AC 10 surf 35/50 mixtures with various Dugny hydrated lime contents. "Theoretical content" refers to the expected value, obtained by correcting the nominal content for the purity of hydrated (Table 5)

Sample	Nominal hydrated lime content in the added filler (%)	Nominal hydrated lime content in the recovered filler (%)	Calcium hydroxide content		
Asphalt Cement including mixed fillers with x% of Dugny lime			Measured content (%)	Theoretical content (%)	Relative deviation from theoretical content (%)
	0	0	0	0	0
	50	23.5	19.7	22.3	11.7
	100	47.0	42.1	44.5	5.4

For the recovered filler, the accuracy of the titration was higher than for laboratory mixed filler: the relative deviation in relation to theoretical content lied between 5% and 12% (Table 7). This could come in part from variations in the percent of filler coming from the aggregate, especially the sand.

International Round Robin Test

The raw test results are presented in a graphical way in Fig. 3. Note that each laboratory was asked to perform two replicates, but some labs performed four, so they then appear a second time (e.g., lab 7 was in this case, so it appears as 7a and 7b). The raw data showed that:

- Labs 3, 24, 32, 33, 34 and 35 had very low lime contents, about half of the mean value. They however measured the right volume V_{eq}. As a matter of fact, these labs calculated half of the correct $Ca(OH)_2$ content due to a confusion between the F factor in the formula (see Section 2.2) and the concentration C of the acid. In other words, F was taken as 0.5 when it should have been 1.

- Lab 16 had very high lime content, about double the mean value. It however measured the right volume V_{eq}. Still, this lab doubled the correct $Ca(OH)_2$ content due again to a confusion between the F factor in the formula (see Section 2.2) and the concentration C of the acid. In other words, F was taken as 2 when it should have been 1.

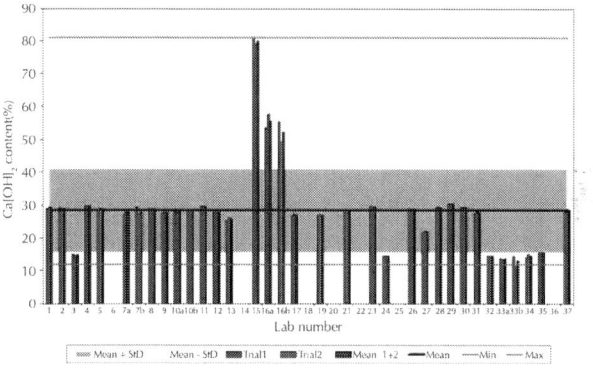

Figure 3: Uncorrected raw results from the round robin test. Note that the nominal lime content was 29.9% in the filler and was not known to the laboratories. Each lab was asked to perform two replicates ("Trial 1" and "Trial 2"), but some labs performed 4 so they then appear a second time (e.g., lab 7 was in this case, so it appears as 7a and 7b). For each lab, the mean of the replicates is given ("Mean 1 + 2"). The mean for all labs is given ("Mean") and the grey area represents the interval for the mean plus or minus one standard deviation. The minimum ("Min") and maximum ("Max") values recorded are also highlighted.

Then, some labs obtained the right result using the wrong acid concentration but involuntarily correcting by making a mistake on the F factor:

- Labs 7, 8, 9, 10 and 11 measured half of the correct V_{eq}, as expected because they involuntarily used an acid at 1 mol/l (instead of 0.5 mol/l as specified in the method) but calculated the correct $Ca(OH)_2$ content by using a F factor of 0.5 instead of 1, repeating a confusion already observed with some laboratories (see above).

- Lab12 used twice the filler content (2 g instead of 1 g as specified in the method) but measured the correct V_{eq} because of the use of a stronger acid (1 mol/l instead of 0.5 mol/l). The final $Ca(OH)_2$ content was still correct because of the same mistake with the F factor.

Since all these mistakes were quite obvious, the data were corrected and the corresponding results are given in Fig. 4. In all cases, this highlights that the presence of the F factor in the formula creates a lot of confusion. Since it brings only limited added value, our proposal would be to simply not mention it in the future and have instead operators make sure that the true concentration of the acid is used.

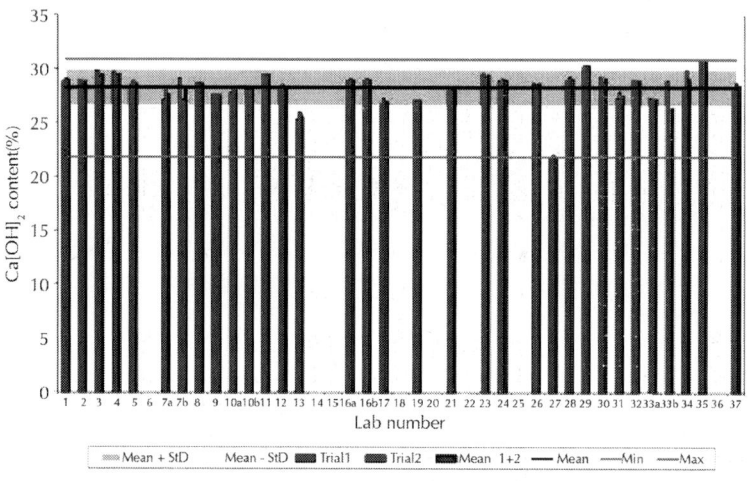

Figure 4: Corrected results from the round robin test. Note that the nominal lime content was 29.9% in the filler and was not known to the laboratories. Each lab was asked to perform two replicates, but some labs performed 4 so they then appear a second time (e.g., lab 7 was in this case, so it appears as 7a and 7b). Anomalous data were corrected as described in the text. For each lab, the mean of the replicates is given ("Mean 1 + 2"). The mean for all labs is given ("Mean") and the grey area represents the interval for the mean plus or minus one standard deviation. The minimum ("Min") and maximum ("Max") values recorded are also highlighted.

Once corrected as explained before, the results gave more consistent data except for lab 15, which was removed from the analysis. As a matter of fact, the testing of lab 15 was done with only 0.01 g of filler instead of 1 g, conditions that are too far from the requirements of the method to be accepted. Such a too small sampling gives very high uncertainties hence the result. This is why the result was removed from the final data set that was used for the statistical analysis (Fig. 4).

Clearly, the range of final values in Fig. 4 was a lot smaller than without corrections (Fig. 3). In the end, the statistical analysis was performed using 33 results from 30 labs. The minimum value was found to be 21.8% and the maximum, 29.9%. The mean for all labs was 28.2% with a standard deviation of 1.6% in absolute terms or 5.8% in relative terms. Note that the nominal lime content was 29.9% in the filler and was not known to the laboratories.

Based on ISO 5725-2 [21], the repeatability was calculated from the standard deviation for within lab variation in the 2 replicates, found to be 0.3% in absolute terms. Therefore, the repeatability r for the quantification was found to be $r = 0.7\%$ in absolute terms on the $Ca(OH)_2$ content in the recovered filler.

Based on ISO 5725-2, the reproducibility was calculated from the standard deviation for between and within lab variation, found to be 1.6% in absolute terms. Therefore, the reproducibility R for the quantification was found to be $R = 4.5\%$ in absolute terms on the $Ca(OH)_2$ content in the recovered filler.

For comparison purposes, it is worth repeating that a previous round robin test was performed in Germany with 12 laboratories [16]. The repeatability (in terms of absolute % weight of hydrated lime in the filler) was $r = 0.52\%$ and the reproducibility was $R = 0.91\%$ for a mean value of 27.3%. The results of this new international round-robbin are therefore consistent in terms of repeatability but show an increased reproducibility. The method remains quite robust anyway.

DISCUSSION

The mean value in the round robin (28.2%) remained significantly below the theoretical one (29.9%). Apart from manufacturing issues (less lime or more filler than wanted could have finally ended up in the mixture), this is consistent with former observations thought to arise from hydrated lime consumption due to chemical interactions with the aggregate (see Fig. 2 – [14]).

This suggests that the use of this method for production control would be better suited if a first calibration was made in order to assess the lime consumption factor for the aggregate used in the asphalt mixture, and then use this calibration factor to express the measured $Ca(OH)_2$ content in terms of initial $Ca(OH)_2$ content, i.e., the quantity of hydrated lime that was present in the mixer/drum before reacting with the aggregate.

In more practical terms, the procedure will consist in first manufacturing a calibration mixture in the lab with a known nominal hydrated lime content similar to the one intended in the jobsite. From this mixture, the filler can be recovered as explained in the "Materials and Procedures" section and $Ca(OH)_2$ content can be assessed with the German method as explained before. Then, the inverse of what was called the recovery rate in Fig. 2[14] could be called the aggregate correction factor ACF defined as:

$$ACF = \text{nominal } Ca(OH)_2 \text{ content/measured } Ca(OH)_2 \text{ content}$$

In the above case, the ACF would therefore be 1.06 (=29.9/28.2). This means that controls could be performed using the German method, and in addition to giving the measured $Ca(OH)_2$ content as detailed above, a calculated hydrated lime content in the mixture could be obtained from the following formula:

Calculated hydrated lime content

¼ **measured** $Ca(OH)_2$ content

ACF = filler content in the mixture

Therefore, the method could provide a way, after calibration in the lab in order to measure the ACF, to directly and accurately express the results in terms of hydrated lime content in the mixture. Note that measuring the filler content in the mixture is a usual control practice and therefore does not represent an additional work. Former work already displayed in Fig. 2, together with the results shown in this paper, show that the ACF can range from 1.06 to 1.67.

Finally, the physico-chemical origin of the ACF remains to be fully understood. Former work performed with asphalt mixtures made with the same bitumen and differing only by aggregate origin (Fig. 2), still showed a wide range of ACF. Therefore, the ACF cannot be attributed to bitumen consumption or filler recovery method. The most likely origin, although still to be fully explained, remains that part of the hydrated lime gets consumed in reactions with the aggregate [13].

CONCLUSIONS

This article first presented the two existing methods found in the literature, one from the USA and the other from Germany, in order to quantify the hydrated lime content in an asphalt mixture. The US method is based on Infra-Red spectroscopy while the German method is based on acid-base titration.

This last one seems to be the easiest to install in control laboratories given that the test set-up, i.e., a titrator, is cheap and simple and already found in most road laboratories. Therefore, it was carefully evaluated, first in one laboratory, then in a European round robin gathering 37 laboratories from 8 different countries. The round robin consisted in quantifying the calcium hydroxide content in the same sample of recovered filler sent to each lab.

The German method consists in first extracting the filler from the mixture using the standard method EN 12697-1 already used daily in asphalt laboratories. The recovered filler is then titrated with a hydrochloric acid solution and the calcium hydroxide content of the tested filler is then obtained. Knowing the filler content in the mix, it is easy to calculate the calcium hydroxide content which is almost similar to the hydrated lime content in the mix.

The method was seen to be quite robust, with a mean value of 28.2%, a repeatability $r = 0.7\%$ and a reproducibility $R = 4.5\%$ in absolute terms. Given that the nominal content was 29.9%, a slight but significant deviation was observed.

This deviation could come from manufacturing issues (less lime or more filler) but could also come from hydrated lime consumption by the aggregate, as observed in a former study [14]. Therefore, the use of this method for systematic control would better be performed after first evaluating the lime consumption factor of the mixture to be controlled, called the Aggregate Correction Factor (ACF). With this preliminary work, control results could be directly expressed in terms of hydrated lime content in the mixture provided the filler content in the mixture is known. Based on this work and a previous study, the ACF can range from 1.06 to 1.67. In all cases, the robustness and simplicity of the method makes it a valuable method for the asphalt producers and the road administrations in order to assess the hydrated lime content in an asphalt mixture. Also, the method was tested in the presence of limestone filler, confirming that the test conditions, especially acid dilution and titration speed, allow quantifying the hydrated lime only and not the carbonates.

Finally, the science behind the ACF remains to be clarified. The suspected reactions between the hydrated lime and the mineral aggregate mentioned in [14] still have to be proven. Additional work is needed in order to validate this last point, and this would bring a very valuable contribution to the current understanding of the mechanisms of hydrated lime modification of asphalt mixtures [19].

ACKNOWLEDGEMENTS

This work was performed under the financial support of the European Lime Association (EuLA) Civil Engineering Task Force. The authors would like to thank EuLA for this support. Also, the authors would like to thank all the laboratories that were involved in the round robin test for their kind contribution.

REFERENCES

1.   Love E. Pavements and roads; their construction and maintenance. New York, NY, USA: Engineering Building Record; 1890.
2.   Kennedy TW. Use of hydrated lime in asphalt paving mixtures. Natl Lime Assoc Bull 1984; 325.
3.   Lesueur D. «Hydrated lime: a proven additive for durable asphalt pavements – critical literature review", Brussels: European Lime Association (EuLA) Ed., 2010, <www.eula.be>.
4.   Hicks RG. Moisture damage in asphalt concrete, NCHRP synthesis of highway practice 175. Washington, District of Columbus, USA: Transportation Research Board; 1991.
5.   Hicks RG, Scholz TV. Life cycle costs for lime in hot mix asphalt, vol. 3, Arlington (Virginia, USA): National Lime Association,

2003   (<http://www.lime.org/LCCA/LCCA_Vol_I.pdf>,   <http://www.lime.org/LCCA/LCCA_Vol_II.pdf>,   <http://www.lime.org/LCCA/LCCA_Vol_III.pdf>).

6.   Little DN, Epps JA. The benefits of hydrated lime in hot mix asphalt, Arlington (Virginia, USA): National Lime Association, 2001 (<http://www.lime.org/ABenefit.pdf>).

7.   Sebaaly PE, Little DN, Epps JA. The benefits of hydrated lime in hot mix asphalt, Arlington (Virginia, USA): National Lime Association, 2006   (<http://   www.lime.org/BENEFITSHYDRATEDLIME2006.pdf>).

8.   Little DN, Petersen JC. Unique effects of hydrated lime filler on the performance-related properties of asphalt cements: physical and chemical interactions revisited. J Mater Civ Eng 2005;17(2):207–18.

9.   Sebaaly PE. Comparison of lime and liquid additives on the moisture damage of hot mix asphalt mixtures, Arlington (Virginia, USA): National Lime Association, 2007 (<http://www.lime.org/MoistureDamageHotMix.pdf>).

10.   Lesueur D, Petit J, Ritter H-J. Increasing the durability of asphalt mixtures through hydrated lime modification: What evidence? Eur Roads Rev 2012;20:48–55.

11.   Raynaud C. L'ajout de chaux hydratée dans les enrobés bitumineux. BTP Matériaux n 22, October 2009. p. 42–43.

12.   Voskuilen JLM, Verhoef PNW. Causes of premature ravelling failure in porous asphalt. In: Proc. RILEM symposium on performance testing and evaluation of bituminous materials, 2003. p. 191–7.

13.   Lesueur D, Petit J, Ritter H-J. The mechanisms of hydrated lime modification of asphalt mixtures: a state-of-the-art review. Road Mater Pavement Des 2013;14:1–16.

14.   Arnold TS, Rozario J, Youtcheff J. New lime test for hot mix asphalt unveiled. Public Roads 70(5), March/April 2007.

15.   Arnold TS, Rozario-Ranasinghe M, Youtcheff J. Determination of lime in hotmix asphalt. Transp Res Rec 2006;1962:113–20.

16.   Schiffner H-M. Test method for determining hydrated lime in asphalt. CementLime-Gypsum International 2003;56(6):76–82.

17. Technische Prüfvorshriften für Gesteinskörnungen im Strassenbau, Teil 3.9, Bestimmung des Calciumhydroxidgehaltes in Mischfüllern, Ausgabe 2008.

18. European Committee for Standardization, EN 459-2: Building Lime. Part. 2: Test Methods, Brussels (Belgium): European Committee for Standardization, 2012.

19. European Committee for Standardization, EN 12697-1: Bituminous Mixtures. Test Methods for Hot Mix Asphalt. Part 1: Soluble Binder Content, Brussels (Belgium): European Committee for Standardization, 2005.

20. Boynton RS. Chemistry and technology of lime and limestone. 2nd ed. New York (NY, USA): Wiley-Interscience; 1980.

21. International Organization for Standardization. ISO 5725-2: accuracy (trueness and precision) of measurement methods and results – Part 2: Basic method for the determination of repeatability and reproducibility of a standard measurement method. Geneva (Switzerland): International Organization for Standardization; 1994.

12

Recycled Tyre Rubber Modified Bitumens for Road Asphalt Mixtures: A Literature Review

Davide Lo Presti*

Nottingham Transportation Engineering Centre, University of Nottingham, Nottingham, UK

ABSTRACT

Nowadays, only a small percentage of waste tyres are being land-filled. The Recycled Tyre Rubber is being used in new tyres, in tyre-derived fuel, in civil engineering applications and products, in moulded rubber products, in agricultural uses, recreational and sports applications and in rubber modified asphalt applications. The benefits of using rubber modified asphalts are being more widely experienced and recognized, and the incorporation of tyres into asphalt is likely to increase. The technology with much different evidence of success demonstrated by roads built in the last 40 years is the rubberised asphalt mixture obtained through the so-called "wet process" which involves the utilisation of the Recycled Tyre Rubber Modified Bitumens (RTR-MBs). Since 1960s, asphalt mixtures produced with RTR-MBs have been used in different parts of the world as solutions for different quality problems and, despite some downsides, in the majority of the cases they have demonstrated to enhance performance of road's pavement. This study reports the results of a literature review upon the existing technologies and specifications related to the production, handling and storage of RTR-MBs and on their current applications within road asphalt mixtures. Furthermore, considering that RTR-MBs technologies are still struggling to be fully adopted worldwide, mainly because of poor information, lack of training of personnel and stakeholders and rare support of local policies, the present work aims to be an up-to-date reference to clarify benefits and issues associated to this family of technologies and to finally provide suggestions for their wide-spread use

TYRE RUBBER: ENVIRONMENTAL PROBLEM OR ENGINEERING RESOURCE?

The increasing number of vehicles on the roads of industrialised and developing nations generates millions of used tyres every year. About 1.4 billion tyres are sold worldwide each year and subsequently as many eventually fall into the category of end of life tyres (ELTs) (Fig. 1). Moreover, the amount of ELTs in Europe, US and Japan are about to increase because of the projected growing number of vehicles and increasing traffic worldwide. These tyres are among the largest and most problematic sources of waste, due to the large volume produced and their durability. The US Environmental Protection Agency reports that 290 million scrap tyres were generated in 2003 (EPA, 2007). Of the 290 million, 45 million of these scrap tyres were used to make automotive and truck tyre re-treads. In Europe every year, 355 million tyres are produced in 90 plants, representing the 24% of world production [1]. In addition the EU has millions of used tyres that have been illegally dumped or stockpiled. The inadequate disposal of tyres may, in some cases, pose a potential threat to human health (fire risk, haven for rodents or other pests such as mosquitoes) and potentially increase environmental risks. Most countries, in Europe and worldwide, have relied on land filling to dispose of used tyres but the limited space and their potential for reuse has led to many countries imposing a ban on this practice. The current estimate for these historic stockpiles through-out the EU stands at 5.5 million tonnes (1.73 times the 2009 annual used tyres arising) and the estimated annual cost for the management of ELTs is estimated at   600 million [2]. With landfills minimising their acceptance of whole tyres and the health and environmental risks of stockpiling tyres, many new markets have been created for scrap tyres.

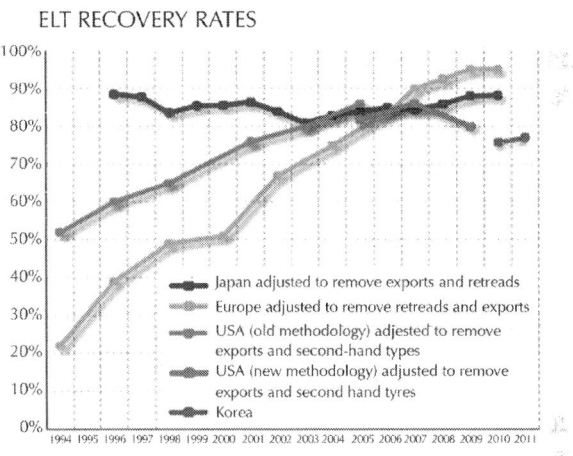

Figure 1: Evolution of ELTs recovery rates in major tyre markets, adapted from [2].

In order to face this problem, in Europe in 1989, a used tyres group composed of experts from the main tyre manufacturers producing in Europe, was set up under the strategic guidance of the European Tyre and Rubber Manufacturers Association (ETRMA). This Group was dedicated to the management of end of life tyres (ELTs). Also thanks to this group, since 1996, the collection rate has increased steadily while there has been a continuous decline in the land filling of used tyres (Table 1). In 2009 the European Union was faced with the challenge of managing, in an environmentally sound manner, more than 3.2 million tonnes of used tyres of which 95% were recovered. This confirms Europe as one of the most active areas in the world in the recovery of ELTs.

Table 1: ELTS management trends in the EU, adapted from [1]

	1994 (%)	1996 (%)	1998 (%)	2000 (%)	2002 (%)	2004 (%)	2006 (%)	2008 (%)	2008 (%)
Re-use/export	11	81	11	10	11	9	9	8	10
Reconstruction	10	12	11	11	11	12	11	10	8
Material recovery	6	11	18	19	25	28	34	38	40
Energy recovery	11	20	20	21	27	31	31	37	38
Landfill (disposal)	62	49	40	39	26	20	13	6	4

Country arise and recovery rates demonstrate that ELTs management in Europe is allowing the progressive elimination of land filling and raises the availability of Recycled Tyre Rubber (RTR) to be recycled for other purposes. In fact, the same characteristics that make waste tyres such a problem also make them one of the most re-used waste materials, as RTR is very resilient and can be reutilised in other products. These efforts should for example help to further develop the use of ELTs in rubberised asphalt in road construction, which has high growth potential in Europe and it is still relatively underutilised [2].

Recycled Tyre Rubber as Engineering Material

The tyre is a complex and high-tech safety product representing a century of manufacturing innovation, which is still on-going. From the material point of view the tyre is made up of three main components materials: (i) elastomeric compound, (ii) fabric and (iii) steel. The fabric and steel form the structural skeleton of the tyre with the rubber forming the "flesh" of the tyre in the tread, side wall, apexes, liner and shoulder wedge [3]. This engineering process is necessary to transform natural rubber in a product able to ensure performance, durability and safety. In fact, natural rubber is sticky in nature and can easily deform when heated up and it is brittle when cooled down (Table 2). In this state it cannot be used to make products with a good level of elasticity. The reason for inelastic deformation of not-vulcanised rubber can be found in the chemical nature as rubber is made of long polymer chains. These polymer chains can move independently relative to each other, and this will result in a change of shape. By the process of vulcanisation cross-links are formed between the polymer chains, so the chains cannot move independently anymore. As a result, when stress is applied the vulcanised rubber will deform, but upon release of the stress the rubber article will go back to its original shape. Compounding is finally used to improve the physical properties of rubber by incorporating the ingredients and ancillary substances necessary for vulcanisation, but also to adjust the hardness and modulus of the vulcanised product to meet the end requirement. Different substances can be added according to the different tyre mixtures; these include mineral oil and reinforcing fillers as carbon black and silica [4]. In general, truck TR contains larger percentages of natural rubber compared to that from car tyres[5]. Table 3 summarises the general tyre composition of tyres used in cars and trucks in the EU [2].

Table 2: Effect of temperature on natural rubber, adapted from [2]

-10 °C	Brittle and opaque
20 °C	Soft, resilient and translucent
50 °C	Plastic and sticky
120 °C -160 °C	Vulcanised when agents e.g., sulphur are added
180 °C	Break down as in the masticator
200 °C	Decomposes

Table 3: Comparison of passenger car and truck tyres in the EV, adapted from [3]

Material (contents)	Car (%)	Truck/buses (%)
Rubber/elastomers	48	43
Carbon black	22	21
Metal	15	27
Textile	5	–
Zinc oxide	1	2
Sulphur	1	1
Additives	8	6

From the structural point of view, the main components of a tyre are the tread, the body, side walls and the beads (Fig. 2). The tread is the raised pattern in contact with the road. The body supports the tread and gives the tyre its specific shape. Beads are metal-wire bundles covered with rubber, which holds the tyre on the wheel. The inherent characteristics of the tyre are the same worldwide. They include: the resistance to mould, mildew, heat and humidity, retardation of bacterial development, resistance to sunlight, ultraviolet rays, some oils, many solvents, acids and other chemicals. Other physical characteristics include their non-biodegradability, non-toxicity, weight, shape and elasticity. However, many of the characteristics, which are beneficial during their on-road life as consumer products, are disadvantageous in their post-consumer life and can create problems for collection, storage and/or disposal [6].

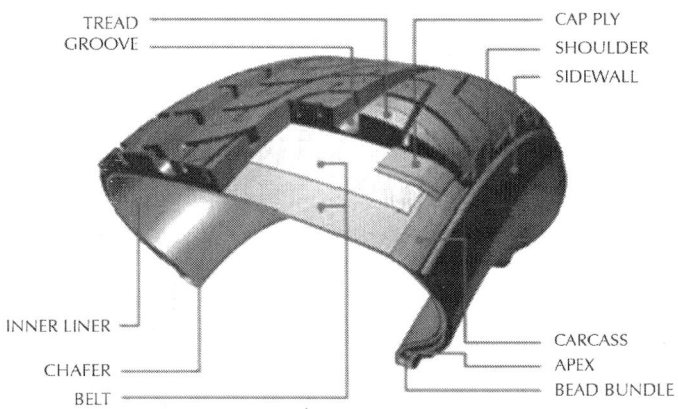

Figure 2: Tyre structure, adapted from [7].

From ELTs to Crumb Rubber Modifier

The tyre life cycle traditionally comprises five main stages, which includes extraction, production, consumption, collection of used tyres and waste management. A simplified version of the tyre cycle is illustrated in Fig. 3. After the collection of ELTs, the next stage includes landfilling and recovery. Worldwide there has been a continuous decline in landfilling used tyres, while the recovery routes include different options such as: "energy recovery" where ELTs having a calorific value equivalent to that of good quality coal are used as an alternative to fossil fuels, or "chemical processing" such as pyrolysis, thermolysis and gasification, (the economic viability of these options has yet to be proved) and finally "granulate recovery". The latter involves tyre shredding and chipping processes which is carried out by using large machinery that cuts up tyres into small pieces of different sizes. At this stage, after the removal of the steel and fabric components, the material RTR can be used for a variety of civil engineering projects: rubberised asphalt pavements, flooring for playgrounds and sports stadiums, as shock absorbing mats, paving blocks, roofing materials, etc.

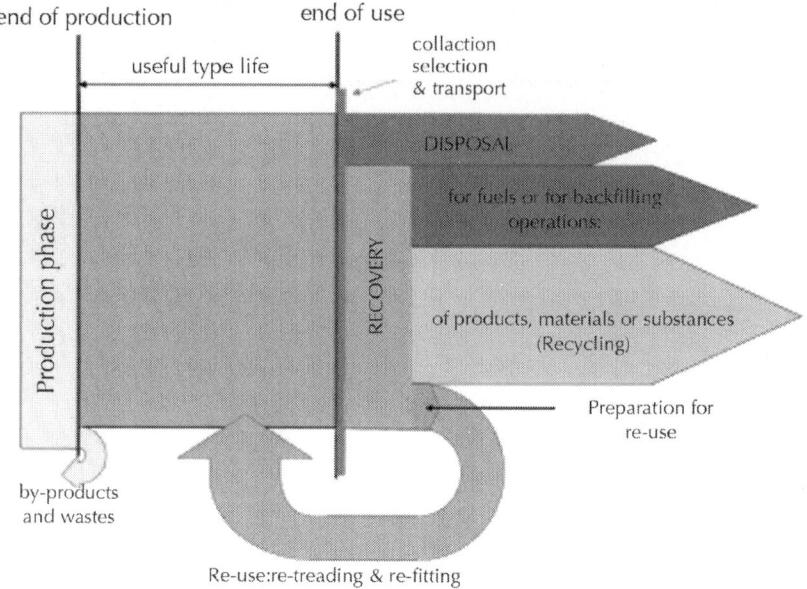

Figure 3: Life cycle of end of life tyres, adapted from [8].

The size of the tyre shreds may range from as large as 460 mm to as small as 25 mm, with most particles within the 100 mm to 200 mm range, while the tyre chips range from 76 mm down to approximately 13 mm. By further reducing the size of shreds and chips, it is possible to produce Ground and Crumb Rubber, also known as size-reduced rubber, which are suitable to be re-used in the asphalt industry. Crumb Rubber Modifier (CRM) is the common name used to identify the RTR particles used to modify bitumen. It is important to recognize that today CRM is typically a highly controlled material. The process is no longer just grinding up a stockpile of ELTs and adding the rubber to hot asphalt. The handling and shredding process is carefully planned and monitored to produce a clean and highly consistent rubber material. During the process, the tyre's reinforcing wire and fiber is removed. The steel is removed by magnets and the fiber is removed by aspiration. The resulting rubber particles are consistently sized and very clean. Automated bagging systems help ensure proper bag weights and eliminate cross contamination [9].

There are several technologies to reduce ELTs in CRM.

Ambient Grinding

This is a method of processing where scrap tyre is ground or processed at or above ordinary room temperature. Ambient processing is typically required to provide irregularly shaped, torn particles with relatively large surface areas to promote interaction with the paving bitumen. This is a mechanical grinding, performed by means of rotating blades and knives, in which the critical step is the separation of the fibers, amongst which are generally included steel fibers. Once separated from the metallic material, ambient grinding is able to produce rubber crumbs with grain size ranging from 5 to 0.5 mm. Ambient grinding is the most commonly used and probably the most cost effective method of processing end of life tyres. A schematic representation of ambient grinding is shown in Fig. 4.

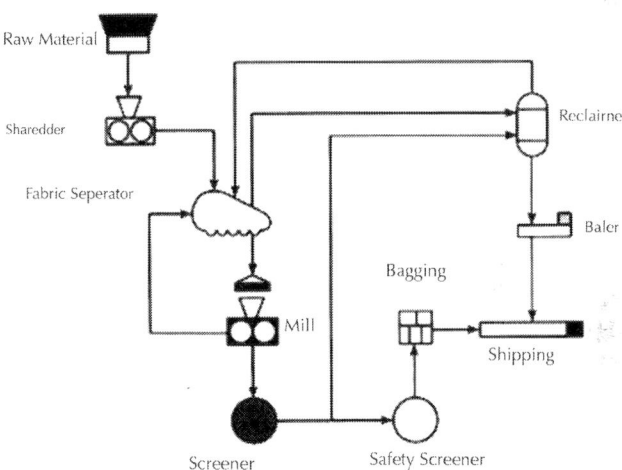

Figure 4: Schematic representation of ambient grinding, adapted from [3].

Cryogenic Grinding

As shown in Fig. 5, this process uses liquid nitrogen to freeze the RTR (typically between -87 to -198 °C) until it becomes brittle, and then uses a hammer mill to shatter the frozen rubber into smooth particles with relatively lower surface area (Fig. 6) than those obtained by ambient grinding (Fig. 7 and Fig. 5). Oliver [10] showed that several characteristics of the rubber granulate determine the elastic properties of the Crumb Rubber and those conferred on the final mix: they enhance with the decrease of specific gravity and particle size, and increase with the higher surface porosity of the granules. In fact, in wet process, rubber particles with a smooth surface, showed reduced reaction with the bitumen and worst the elastic properties of the mixture, if compared with those obtained by using granules with bigger porous surfaces and less specific weight. As a result, the use of CRM from cryogenic process in bituminous mixtures is discouraged [11]. A comparison between the properties of cryogenic and ambient ground rubber is summarised in Table 4.

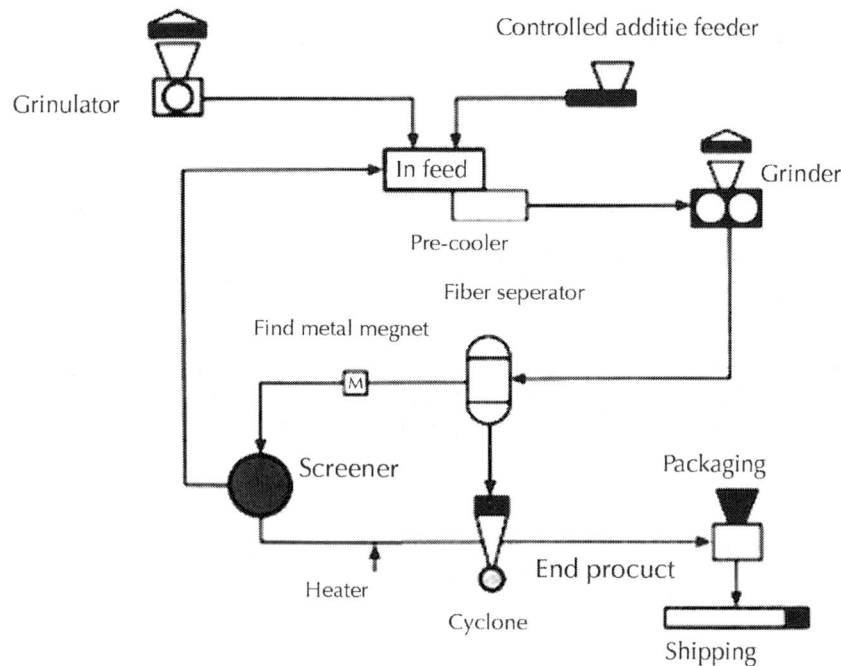

Figure 5: Schematic representation of cryogenic grinding, adapted from [3].

Figure 6: Ambient rubber crumbs. SEM analysis at 200um and 400x magnification, adapted from [12].

Figure 7: Cryogenic rubber crumbs. SEM analysis at 200um and 400x magnification, adapted from [12].

Table 4: Comparison between ambient and cryogenic ground rubber

Physical property	Ambient	Cryogenic
Specific gravity	Same	Same
Particle shape	Irregular	Regular
Fibre content	0.5%	Nil
Steel content	0.1%	Nil
Cost	Comparable	Comparable

Other Processes

In addition to conventional ambient grinding techniques and the cryogenic process, there other less common proprietary processes currently in use to reduce RTR in crumbs or fine powder to be used as CRM:

- *Wet-grinding* is a patented grinding process in which tiny rubber particles are further size reduced by grinding into a liquid medium, usually water. Grinding is performed between two closely spaced grinding wheels. The obtained fine mesh RTR is mainly used as bitumen modifier [12].

- *Hydro jet size reduction*. This is a technique of processing RTR into finer particles with the help of pressurised water. Water jets at very high operating pressure (around 55.000 psi) rotate in high speed arrays producing clean, wire-free rubbers crumbs (Fig. 8). However, the process protocol is relatively new and still unknown to most in the industry. Nevertheless the high level of roughness of the resulting rubber crumbs makes this product very much attractive for bitumen modification. Fig. 8 shows an example of a microscopic analysis of crumbs obtained through a patented hydro jet size reduction process.

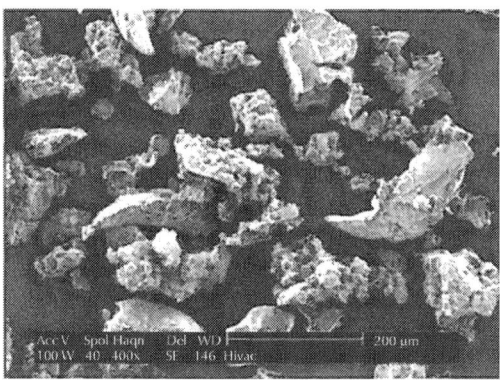

Figure 8: Hydrojet rubber crumbs. SEM analysis at 200um and 400x magnification, adapted from [12].

History of RTR in Road Asphalt Mixtures

The accumulation of ELTs and premature pavement failures are both interconnected and dependant of each other due to enormous increase in traffic density and axle loading respectively. The use of RTR in asphalt pavements started 170 years ago, with an experiment involving natural rubber with bitumen in the 1840s [13], attempting to capture the flexible nature of rubber in a longer lasting paving surface. In 1960s scrap tyres were processed and used as a secondary material in the pavement industry. One application was introduced by two Swedish companies which produced a surface asphalt mixture with the addition of a small quantity of ground rubber from discarded tyres as a substitute for a part of the mineral aggregate in the mixture, in order to obtain asphalt mixture with improved resistance to studded tyres as well as to snow chains, via a process known as "dry process" [14]. In the same period Charles McDonalds, a materials engineer of the city of Phoenix in Arizona (USA), was the first to find that after thoroughly mixing crumbs of RTR with bitumen (CRM) and allowing it to react for a period of 45 min to an hour, this material captured beneficial engineering characteristics of both base ingredients. He called it Asphalt Rubber and the technology is well known as the "wet process" (Fig. 9). By 1975, Crumb Rubber was successfully incorporated into asphalt mixtures and in 1988 a definition for rubberised bitumen was included in the American Society for Testing and Materials (ASTM) D8 and later specified in ASTM D6114-97. In 1992 the patent of the McDonald's process expired and the material is now considered a part of the public domain. Furthermore, in 1991, the United States federal law named "Intermodal Surface Transportation Efficiency Act" (then rescinded), mandated its widespread use, the Asphalt-Rubber technology concept started to make a "quiet come back" [15]. Since then, considerable research has been done worldwide to validate and improve technologies related to rubberised asphalt pavements.

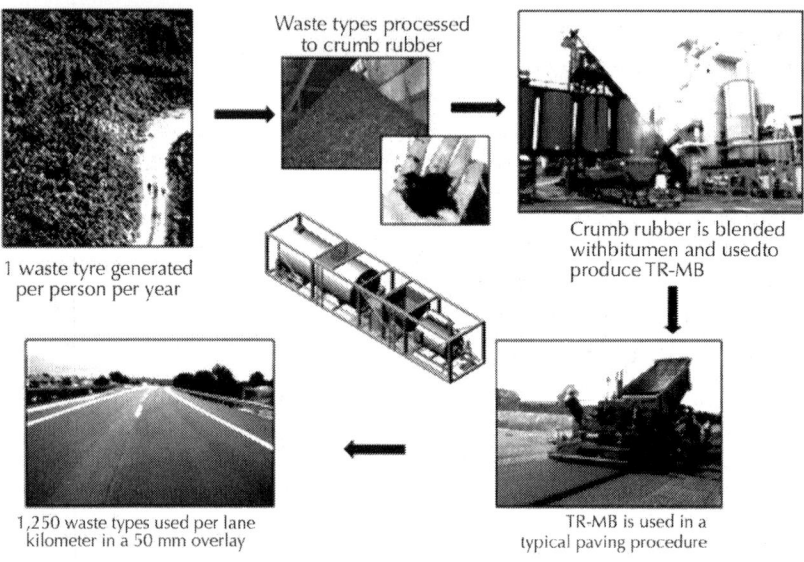

Figure 9: Scheme of rubberised asphalt production through the "wet process".

Nowadays, these rubberised bitumen materials, obtained through the wet process, have spread worldwide as solutions for different quality problems (asphalt binders, pavements, stress absorbing lays and inlayers, roofing materials, etc.) with much different evidence of success demonstrated by roads built in the last 30 years.

RECYCLED TYRE RUBBER MODIFIED BITUMENS (RTR-MBS)

Geography of RTR-MBs in Road Asphalt Mixtures

Since the invention of McDonald, the wet process technology has been used and modified more widely in four states in the US: Arizona, California, Texas, and Florida. More recently wet process has been used also in South Carolina, Nevada and New Mexico. The preference for using this particular modifier was due to the fact that not only does the utilisation of ELTs solve environmental problems but it also offers other benefits such as increased skid resistance, improved flexibility and crack resistance, and reduced traffic noise [16].

South Africa and Australia started introducing bitumen–rubber as a binder for asphalt and for seals from the early 1980s and mid 1970s respectively. In South Africa, both wet and dry processes were reported to have been used successfully although the dry process was mainly used in asphalt [17]. Two states in Australia (New South Wales and Victoria) adopted the wet process for limited application of rubberised asphalt, mainly as a crack resisting layer, but otherwise its usage has been predominantly for sprayed seal applications [18].

In Europe wet rubberised asphalt has been successfully used in road pavements application since 1981 in Belgium, as well as in France, Austria, Netherlands, Poland and Germany [19], more recently also in Greece [20] and UK [21], but the countries with a higher numbers of applications are Portugal [22], Spain [23], Italy [24], Czech Republic [25] and Sweden [26].

Nowadays the rubberised asphalt technology is being adopted in many other parts of the world: As reported by Widyatmoko and Elliot [18], Taiwan was reported to have adopted the Arizona DOT gap-graded and open-graded rubberised asphalt mixtures for flexible pavement rehabilitation; furthermore, rubberised asphalt has been trialled in Beijing and for use in new and maintenance work as part of the preparation for the 2008 Olympics in China and it has also been used in EcoPark Project in Hong Kong. On the basis of first positive experiences also Brazil [27] and Sudan [28] are strongly investing in the application of this technology for road pavements.

Overview of the Bitumen – RTR Interaction Process

The term "wet process" refers to a whole family of technologies which varies a lot with regards with the chosen processing conditions. The nature of the mechanism by which the interaction between bitumen and CRM takes place has not been fully characterised. Traditionally it is reported that bitumen–rubber interaction is not chemical in nature [13], but other studies claim that the increase in binder viscosity cannot be accounted for only by existence of the rubber swelling particles [29].

The reaction itself is made up of two simultaneous processes (Fig. 10): partial digestion of the rubber into the bitumen on one hand and, on the other, adsorption of the aromatic oils available in this latter within the polymeric chains that are the main components of the rubber, both natural and synthetic, contained in the RTR. The absorption of aromatic oils from the bitumen into the rubber's polymer chains causes the rubber to swell and soften [30].

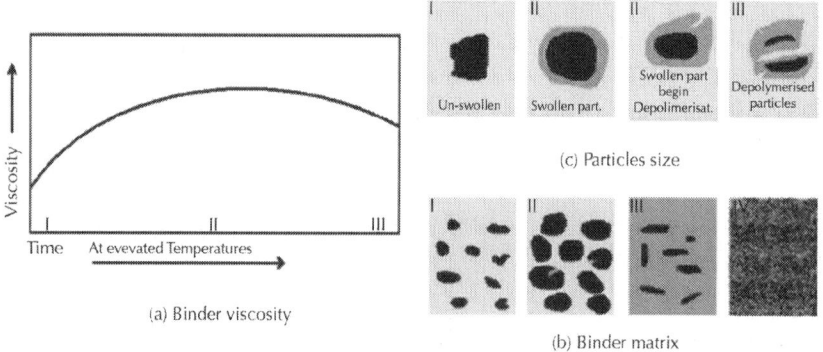

(a) Binder viscosity

(c) Particles size

(b) Binder matrix

Figure 10: Bitumen-RTR interaction phenomenon at elevated temperatures: change of properties over time Adapted from [31].

Rubber particles are swollen by the absorption of the bitumen oily phase at high temperatures (160–220 °C) into the polymer chains, which are the key components of the RTR-MB to form a gel-like material. Therefore, during the reaction there is a contemporaneous reduction in the oily fraction and an increase of rubber particle sizes with a consequent reduction of the inter-particle distance. This implies the formation of gel structures that produce a viscosity increase up to a factor of 10 [13].

Rubber reacts in a time–temperature dependent manner. If the temperature is too high or the time is too long, the swelling will continue to the point where, due to long exposure to the high temperatures, swelling is replaced by depolymerisation/devulcanisation which causes dispersion of the rubber into the bitumen. Depolymerisation starts releasing rubber components back to the liquid phase causing a decrease in the complex modulus (G^*), which is related to the material stiffness, while the the phase angle (δ), related to the elastic properties, continue to modify (Fig. 10a and b). If temperature is high or time is long enough, depolymerisation will continue causing more destruction of the binder networking and so δ modification is lost [32]. The interaction between bitumen and rubber materials is material-specific and depends on a number of basic factors, including:

- Processing variables: temperature, time and device (applied shear stress).
- Base binder properties: bitumen source and eventual use of oil extenders.
- RTR properties: source, processing methods, particle size and content.

These variables represent the processing/interaction conditions that are necessary to monitor during the mixing of rubber within bitumen in order to govern the modification process. RTR-MBs are extremely dependent on the variability of these processing conditions, particularly to what concerns the temperature and time of reaction [33]. Moreover, RTR-MBs must be properly designed and, where necessary, produced to comply with specifications and provide a quality product suitable for the expected climate and traffic conditions. Depending on the adopted processing system, on the chosen processing conditions and on the selected materials, the wet process leads to different technologies as explained in the next sections.

Terminology Associated with RTR-MBs

On the grounds of research done around the globe, rubberised bitumen is used as a general term to identify a group of concepts that incorporate RTR into bituminous binders for paving applications. These terms refer to the uses of RTR, in form of CRM as modifying agent in bitumen, that are different in their mix composition, method of production or preparation and in their physical and structural properties. The method of modifying bitumen with CRM produced from scrap tyres and other components as required before incorporating the binder into the bitumen paving materials, is referred to as the 'wet process'. Wet process is obtainable through two different processing systems.

McDonald Process

This terminology is related to the system of producing RTR-MB with the original wet process proposed by Charles McDonald in the 1960s. The McDonald blend is a Bitumen Rubber blend produced in a blending tank by blending Crumb Rubber and bitumen. This modified binder is then passed to a holding tank, provided with augers to ensure circulation, to allow the reaction of the blend for a sufficient period (generally 45–60 min). The reacted binder is then used for mix production (Fig. 11).

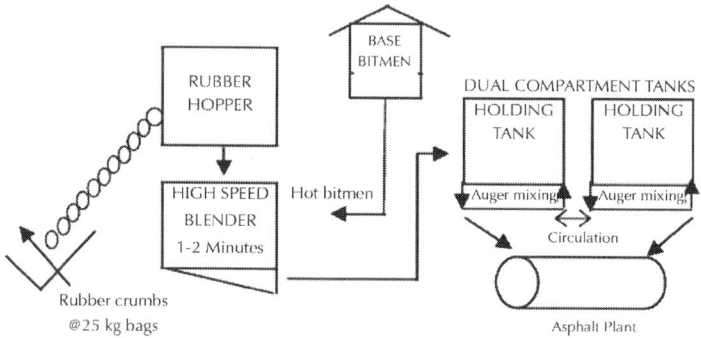

Figure 11: Schematic diagram of McDonald's wet process.

Continuous Blending-Reaction Systems

This system is similar to the McDonald process of blending, the difference is that CRM and bitumen are continuously blended during the mix production or prepared by hand and then stored in storage tanks for later use. Therefore, it consists of a unique unit with agitators, in which the reaction occurs during the blending [13].

Field Blends

Bitumen-RTR blends are typically produced at the asphalt plant by incorporating some modifications to the existing asphalt plant, for this reason they can be identified as field-blends. The above mentioned modifications include the introduction/adaptation of heated blending tanks, heated reaction tanks, rubber feed and storage tanks [13]. Hence, it is already few years that field-blended RTR-MBs can also be blended through less drastic modification to a standard hot mix asphalt plant through the portable production unit based on the McDonald system. The equipment is typically trailer mounted and is transported into the asphalt plant site. The blending unit receives ground rubber in the hopper, the rubber then moves to the mixing chamber to be blended with virgin liquid bitumen (Fig. 12); the resulting rubberised bitumen is stored in the Portable Reaction Tank (Fig. 13). Once reacted, binder is moved to a second compartment where it is fed to the hot-mix plant to be included in the HMA production. At the end, conventional paving equipment without modifications is used to place the material.

Figure 12: Example of mobile Mixing unit for Asphalt-Rubber, adapted from [34].

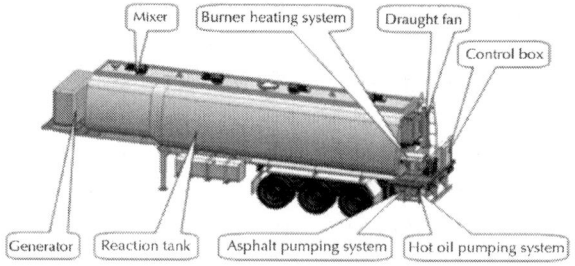

Figure 13: Example of mobile reaction tank for Asphalt-Rubber, adapted from [34].

Terminal Blends

The terminal blended rubberised binder consists of bitumen with Crumb Rubber Modifier (CRM) binder which is digested into the bitumen at the refinery (Fig. 14) or at a bitumen terminal and then delivered to the plant. These binders are blended with various patented processes and they could present problems of phase separation.

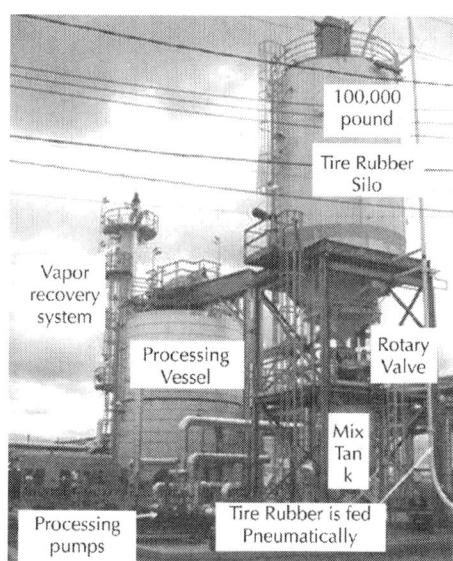

Figure 14: Terminal blend binder system, adapted from [35].

Caltrans Terminology

The most important distinction among the various rubberised bitumens seems to be related to the ability of obtaining homogeneous RTR-MBs which do not present the issue of phase separation during storage periods. Based on this distinction, Caltrans [36] divides the wet process into two families linked to two very different types of RTR modification currently in use: the "wet process-high viscosity" and the "wet process-No-Agitation". To promote a clear understanding, a detailed description of these technologies is provided in the next paragraphs.

WET PROCESS-HIGH VISCOSITY

The original wet process-high viscosity, invented by Charles McDonald, leads to a product with a series of benefits which are basically all linked to the binder's increase in elasticity and viscosity at high temperatures that allows greater film thickness in paving mixes without excessive drain down or bleeding. According to Caltrans [36], the RTR-MB that maintain or exceed the minimum rotational viscosity threshold of 1500 cPs at 177 °C (or 190 °C) over the interaction period should be described as "wet process-high viscosity". These materials require continue agitation, with special equipment, to keep the RTR particles uniformly distributed. They may be manufactured in large stationary tanks or in mobile blending units that pump into agitated stationary or mobile storage tanks to be used directly at the job site. Phase separation is a big issue of these binders that needs therefore to be produced directly on the job-site or rarely transported to the field within tanks equipped with special augers, For this reason, these blends are often well-known as *Field blends*.

Wet process-high viscosity binders typically require at least 15% CRM to achieve the threshold viscosity. However, for some specifications [36] the viscosity requirements are meet also with less than 15% of RTR content. A number of RTR-MBs and mix specifications have been developed based on this original idea and each of them is related to a specific technology. This section will describe each of these technologies listed in Table 5 by providing a step by step framework to operate an appropriate plan production and storage of these materials.

At last, this paragraph will provide an overview of the benefits and limitations related to the wet process-high viscosity.

Table 5: Existing specifications for rubberised asphalt mixtures

Technologies	Specifications	Country
Asphalt-Rubber binder	ASTM D6114/D6114 M, [4] Caltrans' Asphalt-Rubber user guide, [5]	USA
Bitumen Rubber binder	SABITA Manual 19, [6] AsAc Technical Guideline, [7]	RSA
Crumb Rubber binder	Austroads, [8]	AUS and NZ

Asphalt-Rubber Binder (USA)

According to the ASTM D6114 [37], Asphalt-Rubber is "a blend of asphalt cement, CRM and certain additives in which the rubber component is at least 15% by weight of the total blend and has reacted in the hot asphalt cement sufficiently to cause swelling of the rubber particles".

CRM Requirements

According to ASTM D6114, in order to produce Asphalt-Rubber, the rubber should have the following characteristics:

- Less than 0.75% moisture and free flowing.
- Specific gravity of 1.15 ± 0.05.
- No visible nonferrous metal particles.
- No more than 0.01% ferrous metal particles by weight.
- Fibre content shall not exceed 0.5% by weight (for hot mix binder applications).
- Recommends all rubber particles pass the No. 8 (2.36 mm) sieve.
- (Note that Rubber gradation may affect the physical properties and performance of Bitumen Rubber hot mix).

Caltrans (2006) specifies that CRM to be used as modifier must include 25 ± 2% by mass of high natural rubber and 75 ± 2% of CRM from scrap tyre. Both types of rubber must meet specific chemical and physical requirements including gradation and limits on fabric and wire contaminants. The RTR consists primarily of 2 mm to 600 µn sized particles (No. 10 to No. 30 sieve sizes). The high natural rubber CRM is somewhat finer, mostly 1.18 mm to 300 urn sieve sizes (Table 6).

Table 6: Typical gradation of the RTR to be used in the Asphalt-Rubber binder, adapted from [38]

Sieve #	Nominal size (mm)	% Passing
10	2.36	100
16	1.18	75–100
30	600 µm	25–100
50	300 µm	0–45
100	150 µm	0–10
300	75 µm	0

Base Binder Requirements

ASTM D6114 specifies three different types of Asphalt-Rubber. Each of them is associated to suitable bitumen to be used as base of modification

- Type-I binders typically include stiffer base bitumen and are generally recommended for hot climates, such as: AC-20, AR-8000 and PG64-16.

- Type-II binders typically include base bitumen softer than Type-I and are generally recommended for moderate climates, such as: AC-10, AR-4000 and PG58-22.

- Type-III binders typically include the softest grade base bitumen and are generally recommended for cold climates, such as: AC-5, AR-2000 and PG52-28.

where AC (asphalt cement) and AR (Aged Residue) are referred to the American grading systems based on viscosity. For example, an AC 20 asphalt has a viscosity of 2000 poise (+20%) at 60 °C, whilst an AR 4000 bitumen has a viscosity of 4000 poise (+25%) at 60 °C after aging. A rough comparison between penetration and viscosity (AC and AR) bitumen grades is shown in Fig. 15.

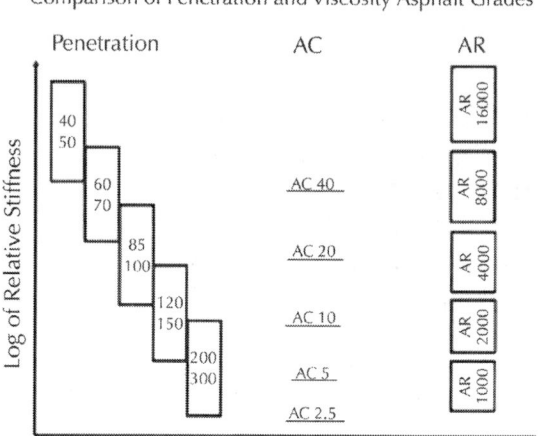

Figure 15: Comparison of Penetration and Viscosity bitumen grades.

Asphalt-Rubber Plant Production

By definition, Asphalt-Rubber is prepared using the "wet process-high viscosity" system. Physical property requirements are listed in ASTM D 6114, "Standard Specification for Asphalt-Rubber binder" (Table 8). The Asphalt-Rubber is produced at elevated temperatures (≥ 350 F, 177 °C) in low shear (Shatanawi, 2010) to promote the physical interaction of the asphalt binder and rubber constituents, and to keep the rubber particles suspended in the blend. Various petroleum distillates or extender oil may be added, at a rate of 2.5–6% by mass of the bitumen binder, to reduce viscosity, facilitate spray applications, and promote workability. Field production of high viscosity Asphalt-Rubber binders is a relatively straightforward process and it is still based on the McDonald process (Fig. 11). Equipment for feeding and blending may differ among Asphalt-Rubber types and manufacturers, but the processes are similar. The component materials are metered into high shear blending units to incorporate the correct proportions of extender oil and CRM into the paving grade asphalt. The blending units thoroughly mix the CRM into the hot asphalt cement and extender oil, and the blend is pumped into a heated tank where the Asphalt-Rubber interaction proceeds.

RTR is usually supplied in 1 ton (0.91 tonne) super sacks that are fed into a weigh hopper for proportioning. Augers are needed to agitate the Asphalt-Rubber inside the tanks to keep the CRM particles well dispersed, otherwise the particles tend either to settle to the bottom or float near the surface. Agitation can be verified by periodic observation through the top hatch or the port where the auger control is inserted. The Asphalt-Rubber binder must be interacted with agitation for a minimum of 45 min at temperatures from 190 to 218 °C to achieve the desired interaction between bitumen and rubber. In order to maintain the reaction temperature within the specified range, the bitumen must be hot, 204–224 °C before the design proportions of scrap tyre CRM and high natural rubber CRM are added. This is because the CRM is added at ambient temperature (not heated) and reduces the temperature of the blend. The component materials are metered into blending units to incorporate the correct proportions of CRM into the bitumen, and are thoroughly mixed. The Asphalt-Rubber producer is allowed to add the extender oil while adding the rubber, although in some cases the base binder may be supplied with the extender included. If the Asphalt-Rubber producer adds the extender oil, use of a second meter is recommended to best control the proportioning. The meter for the extender oil should be linked to that for the bitumen. An Asphalt-Rubber binder interacted at lower temperatures will never achieve the same physical properties as the laboratory design, although it may achieve the minimum specification values for use. Hand held rotational viscometers are used to monitor the viscosity of the Asphalt-Rubber interaction over time for quality control and assurance. Before any Asphalt-Rubber binder can be used, compliance with the minimum viscosity requirement must be verified using an approved rotational viscometer (e.g. Brookfield). As long as the viscosity is in compliance and the interaction has proceeded for at least 45 min, the Asphalt-Rubber may be used [36].

Caltrans (2006) recognises that before starting the plant production, an appropriate Asphalt-Rubber binder design must be developed (Table 7). It also specifies that at least 2 weeks prior to start of construction the Contractor must supply to the Engineer, for approval, an Asphalt-Rubber binder formulation (design or "recipe") that includes results of specified physical property tests, along with samples of all of the component materials. Samples of the prepared Asphalt-Rubber binder must also be submitted to the Engineer at least 2 weeks before it is scheduled for use on the project.

Table 7: Caltrans specification for rubberised bitumen [36]

Test parameter	Specification limits
Apparent viscosity, Haake, 190 °C: cp	1500–4000
Cone penetration at 25 °C (ASTM D217): dmm	25–70
Softening point (ASTM D36): °C	52–74
Resilience at 25 °C (ASTM D3407): %	Minimum 18

Table 8: Specification for Asphalt-Rubber [37]

Binder specification	ASTM D6114 (2009)		
	Type I	Type II	Type III
Apparent viscosity 177.5 °C (ASTM D 2196): cp	1500–5000	1500–5000	1500–5000
Penetration at 25 °C (ASTM D5): dmm	25–75	25–75	50–100
Penetration at 4 °C (ASTM D5): dmm	Min 10	Min 15	Min 25
Softening point (ASTM D36): °C	Min 57.2	Min 54.4	Min 51.7
Resilience at 25 °C (ASTM D5329): %	Min 25	Min 20	Min 10
Flash point (ASTM D93): °C	Min 232.2	Min 232.2	Min 232.2
After TFOT (ASTM D1754), residual penetration at 4 °C (ASTM D5): %	Min 75	Min 75	Min 75
Climatic region	Hot	Moderate	Cold
Average minimum monthly temperature: °C	Min −1	Min −9	Min −9

Average maximum monthly temperature: °C	Min 43	Max 43	Max 27

Asphalt-Rubber Storage

Caltrans requires heating to be discontinued if Asphalt-Rubber material is not used within 4 h after the 45-min reaction period. The rate of cooling in an insulated tank varies, but reheating is required if the temperature drops below 190 °C. A reheat cycle is defined as any time an Asphalt-Rubber binder cools below and is reheated to 190–218 °C. Caltrans allows two reheat cycles, but the Asphalt-Rubber binder must continue to meet all requirements, including the minimum viscosity.

Sometimes the binder must be held overnight. The bitumen and rubber will continue to interact at least as long as the Asphalt-Rubber remains liquid. The rubber breaks down (is digested) over time, which reduces viscosity. Up to 10% more CRM by binder mass can be added to restore the viscosity to specified levels. The resulting Asphalt-Rubber blend must be interacted at 190–218 °C for 45 min and must meet the minimum viscosity requirement before it can be used.

Bitumen Rubber Binder (RSA)

Bitumen Rubber binder is a terminology used for the rubberised bitumen obtained by the wet process in South Africa. Bitumen Rubber binders combine rubber crumbs (Table 9) with bitumen at high temperatures to achieve a complex two phase product, named non-homogeneous binder. In the Technical Guideline of the South African Asphalt Acadamey [38], is also reported that properties of the modified binder used in hot mix asphalt will influence the engineering properties of the resultant mix, therefore the substitution of a conventional bitumen with a modified binder can result in higher air voids due to reduced workability of the higher viscosity modified binders. For this reason, it is important that split samples of the modified binder are sent to all the participating laboratories (e.g.: supplier, site and control laboratories) and tested before commencement of a project to ensure that the results are within the reproducible limits as prescribed by the test methods.

Table 9: Rubber crumbs requirements for bitumen–rubber, adapted from [38]

Property	Requirement	Test method
Sieve analysis (% mass)		MB-14
Passing screen (mm)		
1.180	100	
0.600	40–70	
0.075	0–5	
Poly-isoprene content (%m/m total hydrocarbon)	25 min	Thermo gravimetric Analysis
Fibre length (mm)	6 max	
Bulk density (g/cm^3)	1.10–1.25	MB-16

Rubber Requirements

In South Africa, the Committee of Land Transport Officials (COLTO)'s specification requires the reclaimed rubber to have not less than 30% natural rubber by mass of hydrocarbon content [39], whilst the SABITA Manual [40] and the Technical Guideline of the South African Asphalt Acadamey [38] specify 60–75% natural rubber by mass of hydrocarbon content, with all rubber particles passing the 1.18 mm sieve. The CRM is produced by a mechanical size reduction process. CRM produced by cryogenic-mechanical techniques are not permitted in South Africa. The CRM must be pulverised, free of fabric, steel cords and other contaminants.

Base Bitumen Requirements

The binder used in the production of the bitumen–rubber must be SABITA B12 or B8 road-grade bitumen (60/70 or 80/100 penetration-grade bitumen respectively) or a blend of these grades to provide a product with a particular viscosity and other prescribed properties.

Bitumen Rubber Production

The binder is manufactured from blending penetration grade bitumen (72–82% by mass), plus extender oil (0–4%) plus rubber crumbs (18–24%) in a patented high shear mixer with a speed of 3000 r.p.m, at a temperature in excess of 180 °C but not more than 220 °C and for short periods before the introduction of rubber. During the addition of the rubber component, the blend cools down considerably and has to be re-heated to a temperature of 190–200 °C to ensure proper digestion of the rubber in the bitumen. Special manufacturing equipment is required to manufacture this highly viscous material. The product has a limited useable life of 4–6 h and, therefore, manufacture usually takes place onsite, or very close to the construction site. Bitumen Rubber binder can be used for surface dressing operations, in which case it is applied with binder distributors specially designed to handle this highly viscous binder. For surface dressing applications, the bitumen–rubber binder is manufactured using the "wet process", which is also the most used for the manufacture of bitumen–rubber hot-mix asphalt. The blending unit consists of a small tank equipped with a high speed stirring device that ensures proper "wetting" of the rubber by the binder and prevents the formation of rubber lumps in the final product. From the blending tank the product is transferred to a digestion tank which could also be a specialised binder distributor. In the digestion tank the product is continually agitated while being heated to the final temperature.

The ratio of components varies depending on the bitumen source, the climatic conditions and the application. The more reliable manufacturers in South Africa nowadays prefer to standardise on 20% rubber content and also to preselect the type of tyres for the modification process. Following the addition of the rubber, a digestion period is required for the rubber to swell and partially dissolve in the bitumen/extender oil blend. The rubber never completely dissolves in the bitumen and the product is thus classed as a non-homogenous binder. The Bitumen Rubber blend is then circulated in a holding tank and heated at high temperatures (190–210 °C) to facilitate the chemical digestion process.

The extender oil could either be added to the penetration grade bitumen before delivery or to the bitumen on site. The requirements of the extender oil are such that it should have a flash point of greater than 180 °C and the percentage by mass of aromatic unsaturated hydrocarbons be greater than 55. To prevent sticking of rubber particles, also an addition of calcium carbonate or talc up to 4% by mass of rubber is permitted. On completion of the digestion period, a hand-held viscometer is used to perform a viscosity test on the product to confirm that sufficient digestion has taken place (Table 10). If approved, the product is ready for application[38].

Table 10: Properties of Bitumen Rubber surfacing seals and asphalt, adapted swfrom [38]

Property		Unit	Test method	Class	
				S-R1	A-R1
Softening point[1]	°C	MB-17	55–62	55–65	
Dynamic viscosity @ 190 °C	dPa s	MB-13	20–40	20–50	
Compression and recovery	5 min		MB-11	>70	>80
	1 h	%		>70	>80
	4 days			>25	n/a
Resilience @ 25		%	MB-10	13–35	13–40
Flow		Min	MB-12	15–70	10–50

Bitumen Rubber Storage

Bitumen Rubber degrades rapidly at application temperatures which are in excess of 200 °C. Therefore the blending of Bitumen Rubber generally takes place in close proximity to the spray site or asphalt mixing plant. On completion of the digestion period, the product generally has a further useable life at the application temperature of approximately 4 h. The rate of degradation will vary depending mainly on the application temperature and can be monitored on-site with a hand held viscometer. The manufacturer of the Bitumen Rubber should supply temperature curves showing the changes in the properties over time.

Fig. 16 shows typical changes in the viscosity properties of a Bitumen Rubber at different temperatures over time. Only sufficient quantities of Bitumen Rubber should be blended at any time in accordance with what can be sprayed or mixed within the application viscosity window of the product. Allowance must be made for changing weather conditions and construction delays. Proper planning and close cooperation between the supplier and contractor is essential to limit the over production of Bitumen Rubber which may result in unnecessary degradation of the Bitumen Rubber over prolonged periods of heating. Product must not be superheated if it is not going to be used. This will enable the product to be superheated at a later stage for reuse if it is still within specification. If it happens to be out of specification then only 25% of the product can be re-blended with new bitumen and RTR. Table 11 shows the recommended temperatures and time limits for the short-term handling, storing, spraying, mixing and application binders modified with Bitumen Rubber. S-R1 is a Surface Seal, A-R1 is for hot mix while C-R1 is a Crack sealant and all of them are Bitumen Rubber binders.

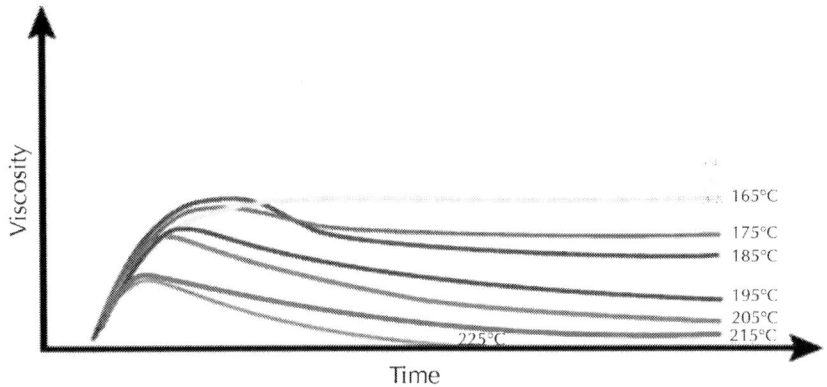

Figure 16: Typical changes in viscosity values from Bitumen Rubber at different temperatures over time, adapted from [38].

Table 11: Typical temperature/time limits for Bitumen Rubber, adapted from [38]

Binder class	Short term handling		Storage		Spraying/asphalt mixing/ application		
	Max temp (°C)	Max holding time (h)	Max temp (°C)	Max holding time (h)	Max temp (°C)	Min temp (°C)	Max holding time (h)
S-R1	165	24	150	240	210	195	Refer to time/ viscosity curve
A-R1	165	24	150	240	210	190	
C-R1	165	24	–	–	190	180	–

Crumb Rubber Modified Binder (AUS & NZ)

Bitumen technologists in Australia have been looking into using rubber in bitumen since the early 1950s, when a surface dressing trial utilising 5% vulcanised reclaimed rubber in R90 bitumen (100 pen) with 10 mm aggregate was applied on the Prince Highway at St. Peters in New South Wales. However, there was not much development until the mid-1970s, when the Road authority of the state of Victoria (VicRoads) in conjunction with the Australian Road Research Board (ARRB) started using Crumb Rubber Modified bitumen in spray seal in urban and rural applications. It has been used extensively as a bitumen modifier in sprayed bituminous seals and occasionally in bitumen applications in Victoria. Thanks to this experience VicRoads, in collaboration with the Roads and Traffic Authority (RTA), New South Wales and the Main roads by Western Australia, have prepared the "Scrap rubber bitumen guide".

Since 1986, success has been reported using rubberised bitumen in Australia, mainly RTA and VicRoads, specifically when used as a crack resisting layer, e.g. thin bitumen overlay over concrete, Stress Absorbing Membranes or Stress Absorbing Membranes Interlayers. In 2006, Austroads published technical specifications framework AP-T41 coupled with a guide to the selection and use of Polymer Modified Binders: AP-T42 including the use of rubberised bitumen, identified as Crumb Rubber Modifed binder [41]. The reports suggest a wide usage of rubber into asphalt with both wet and dry process. The AP-T42 guide includes also a "Crumb Rubber Protocol" in which Constituent Specification, Binder Design, Field Blending, Binder Specification Limits and other restrictions regarding and Sampling and Testing to check requirements are indicated. An important component of this Crumb Rubber specification is the role of laboratory testing as explained later.

CRM Requirements

In Austroads (2006), two levels of rubber content, nominally 15 and 18 are specified for Crumb Rubber Modified binder obtained through the wet process. Another level, from 25 to 30% , is designed for the dry process. Crumbs have only 2 sizes, namely 16 (coarse) and 30 (fine). Bulk density (max 350 kg/mc), grading, maximum particle size, steel content (<0.1%), moisture (<1%) and other properties must be specified by the contractor. All CRM shall be less than 3 mm in length and the rubber shall not contain any foreign material such as sand, fibre or aggregate. All this info is summarised in Table 12.

Table 12: Austroads specification for Crumb Rubber for bitumen modification, adapted from [41]

Test	Methods	Size 16	Size 30
Grading			
Passing 2.36 mm	AG/PT 1:43	100	100
Passing 1.18 mm	AG/PT 1:43	80 (min)	100
Passing 0.60 mm	AG/PT 1:43	10 (max)	40 (max)
Passing 0.15 mm	AG/PT 1:43		5 (max)
Bulk density	AG/PT 1:44	350 kg/m³	350 kg/m³
Water content	AG/PT 1:43	<1%	<1%
Steel content	AG/PT 1:43	<0.1%	<0.1%

Base Bitumen Requirements

The only requirements related to the most suitable bitumen to be used as base for the Crumb Rubber Modified binders, is indicated in the "Crumb Rubber Protocol" of the technical guide of Austroads [41], which specifies to use Bitumen Class 170 corresponding to a Penetration grade of 85/100 (Table 13).

Table 13: Summary of the most important specification requirements for RTR-MBs

Properties	ASTM 2009			Caltrans 2006	AsAc 2007	Austroads 2007
	Type I	Type II	Type III			
Base bitumen requirements						
Penetration: dmm	85–100	120–150	200–300	120–150	60–100	85–100
Crumb Rubber requirements						
Passing sieve: mm	2.36			2.36	1.18	2.36
Rubber content: %	≥15			18–22	18–24	15–18
Additives						
Extender oils: %	–			2.5–6	0–4	–
Caclium carbonate / Talc: %	0–4			–	0–4	–
Processing conditions						
Temperature: °C	177			190–220	180–220	195
Mixing speed: rpm	–			–	3000 rpm	–
Mixing time: min	45 + reaction			45–60	–	30–45

Crumb Rubber Modified Binder Production

As mentioned previously, the Austroads guide for Selection and use of Polymer Modified Binders AP-T42 [41], gives a major importance to the a priori laboratory testing. In fact, the design process requires a laboratory exercise to be undertaken using the following procedure:

1. A trial mix is prepared for each of an appropriate series of rubber concentrations such that the specification limits for the selected binder grade are met in at least one of the mixes (mixing temperature 195 °C, digestion period 45 min for size 16 rubber and 30 min for size 30).

2. The measured properties are plotted against rubber concentration and the concentration at which the specification limits are complied with is identified. This rubber concentration is deemed to be the design concentration.

3. A rubber extraction is performed on the design mix using AG:PT/T1 42 Laboratory method for determining Crumb Rubber concentration. The determined concentration of rubber is used as a check for the analysis of field collected samples.

The diagram in Fig. 17 illustrates the key steps. In the Austroads 2006 specifications, S40R, S45R/S50R and S55R were replaced by S15RF, S18RF, for asphalt application, and S55R was kept for sprayed sealing. Plant protocols may require different procedures to the field defined system and the contractor will be responsible for the binder design. Crumb Rubber Modified binders are produced with production routes: factory and field produced rubberised bitumen.

Factory Produced Crumb Rubber Modified Binders: usually include combining oil and lower concentrations of rubber than their field produced counterpart. Factory produced Crumb Rubber Modified binder are used for hot mix asphalt by using the wet process. This has been shown to improve the mix properties but requires relatively high binder contents. Since the concentration of rubber is an important performance controlling factor, generally the higher the volumetric concentration, the more effective will be the binder in more demanding applications. Flux oil should only be added to the CRM classes and the manufacturer's recommendations should be followed. It is important to use the finer CRM in order to achieve a reasonable level of digestion in the binder in the short time that the material remains hot.

Field Produced Crumb Rubber Modified Binders: are produced for sprayed sealing. They are a high temperature blend of bitumen and CRM without combining oils, and may contain up to 25% by mass of rubber; these formulations are not suitable for long distance transportation or extended storage but represent the highest performing materials within their class when optimally digested. The properties of the field produced rubberised bitumen are generally controlled by their softening point, elastic recovery and/or torsional recovery [41].

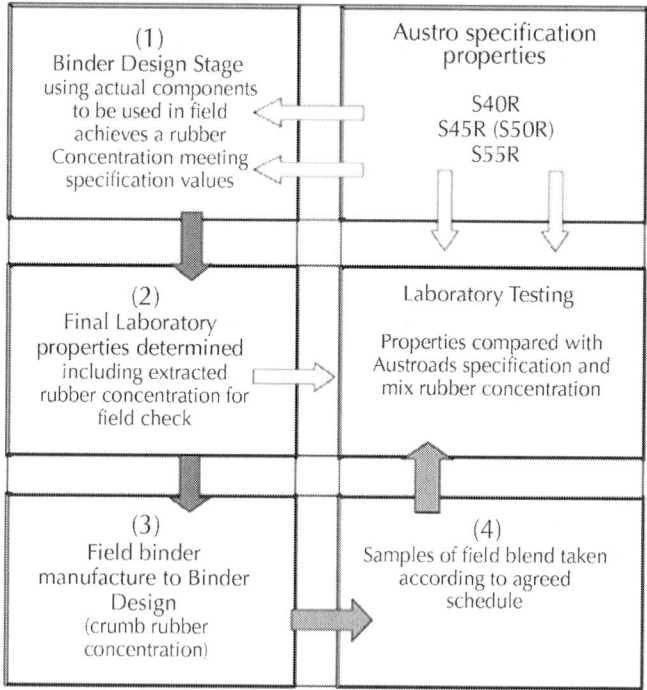

Figure 17: Austroads laboratory procedure for Crumb Rubber modified binders, adapted from [41].

Crumb Rubber Modified Binder Storage

When field mixing Crumb Rubber binders, the material must be continually circulated to minimise settling out of any rubber particles. Failure to do this will most likely result in blockages of the spraying jets and/or pipework. Do not store field produced Crumb Rubber mixtures in bitumen sprayers, road tankers or bulk storage because of the potential problem of segregation and settling out of the rubber particles resulting in blocked pipework etc.

It is worth to highlight that current work in Australia is focussing on the use of Crumb Rubber as a substitute for other modified binders in more conventional applications while using similar design criteria, with particular interest in the use of pre-blended products which behave much more like other pre-blended modified binders. Laboratory investigations with pre-blended Crumb Rubber Modified binders indicate satisfactory performance but at higher binder contents (>8%).

Table 13 summarises some of the most important specifications requirements for the RTR-MB production around the world.

Wet Process-High Viscosity: Benefits, Issues and Limitation

Extensive literature clearly shows the numerous successes obtained using rubberised bituminous mixtures produced with the wet process-high viscosity. However, this technology is not a panacea. This section was thought to explain the main benefits provided by the usage of this technology and the issues that still exist and, in some cases, limit its extensive application.

Benefits

The primary reason for using RTR-MBs is that it provides significantly improved engineering properties over conventional paving grade bitumen. As for Asphalt-Rubber binders, they can be engineered to perform in any type of climate as indicated in ASTM D 6114. At intermediate and high temperatures, the RTR-MB shows physical and rheological properties significantly different than those of neat paving grade bitumens. The rubber stiffens the binder and increases elasticity (proportion of deformation that is recoverable) over these pavement operating temperature ranges, which decreases pavement temperature susceptibility and improves resistance to permanent deformation (rutting) and fatigue with little effect on cold temperature properties [36]. As demonstrated by various researchers, RTR-MBs obtained through the wet process have reduced fatigue and reflection cracking, greater resistance to rutting, improved aging and oxidation resistance and better chip retention due to thicker binder films [42], [43], [44], [45], [39] and [46]. Also Asphalt-Rubber pavements have been demonstrated to have lower maintenance costs [47], lower noise generation [48], [49] and [50] higher skid resistance and better night-time visibility due to contrast in the pavement and stripping [51]. A summary of the benefits of using High-viscosity RTR-MBs for road pavement applications is given below:

High Viscosity RTR-MBs have:

- Increased viscosity that allows greater film thickness in paving mixes without excessive drain down or bleeding.
- Increased elasticity and resilience at high temperatures.
- High Viscosity RTR-MBs pavements have:
- Improved durability.
- Improved resistance to surface initiated and fatigue/reflection cracking due to higher binder contents and elasticity.
- Reduced temperature susceptibility.
- Improved aging and oxidation resistance due to higher binder contents, thicker binder films, and anti-oxidants in the RTR.
- Improved resistance to rutting (permanent deformation) due to higher viscosity, softening points and resilience (stiffer, more elastic binder at high temperatures).
- Lower pavement maintenance costs due to improved pavement durability and performance.
- In addition, High Viscosity RTR-MBs pavements and binders can result in:
- Reduced construction times due to thinner lifts.
- Reduced traffic noise (primarily tyre noise).
- Improved safety due to better long-term colour contrast for pavement markings because carbon black in the rubber acts as a pigment that keeps the pavement blacker longer.
- Savings in energy and natural resources by using waste products and not contributing to the stockpiles.

Limitations

Wet process-high viscosity materials are useful, but they are not the solution to all pavement problems. The Asphalt-Rubber materials must be properly selected, designed, produced, and constructed to provide the desired improvements to pavement performance. Pavement structure and drainage must also be adequate. Limitations on the use of Asphalt-Rubber include [36]:

- High Viscosity RTR-MBs are not best suited for use in dense-graded HMA. There is not enough void space in the dense-graded aggregate matrix to accommodate sufficient rubberised binder content to enhance performance of dense-graded mixes enough to justify the added cost of the binder.

- Construction may be more challenging, as temperature requirements are more critical. Asphalt mixtures with High Viscosity RTR-MB must be compacted at higher temperatures than dense-graded HMA because, like polymers, rubber stiffens the binders at high temperatures. Also, coarse gap-graded mixtures may be more resistant to compaction due to the stone nature of the aggregate structure.

- Potential odour, also if it seems to not be harmful (see environmental issues).

- It is not possible to store High Viscosity RTR-MB at elevated temperatures without equipping storage tanks with augers or paddles. Furthermore,

- these binders cannot be stored for prolonged period. If work is delayed more than 48 h after blending the High Viscosity RTR-MB may not be usable. Takallou and Sainton [52], said "the rubberised binder must be used within hours of its production". The reason is that if over-processed the CRM will be digested to such an extent that it is not possible to achieve the minimum specified viscosity even if more CRM is added in accordance with specified limits [37].

Economic Issues related to the usage of RTR-MBs have been adequately addressed by a large number of research projects and reports and also through long standing construction evaluation. For the sake of brevity, the most notable are discussed below:

Higher initial costs. Costs are higher than conventional bitumen per unit ton until economies of scale are in place. From a study conducted in the USA by Hicks and Epps [53], it has been experienced that the Asphalt-Rubber hot mix could costs almost double than conventional mixes. Nevertheless, since the year 2000 a falling cost difference trend has been registered, also because costs of construction materials and petroleum products have increased (Fig. 18). Given this change in the cost structure it is easy to observe that High Viscosity RTR-MB is presently very attractive in cost when particularly examined in light of actual usage. With the RTR being a very cheap waste material the RTR-MBs technologies are increasing their appeal also from the economical point of view.

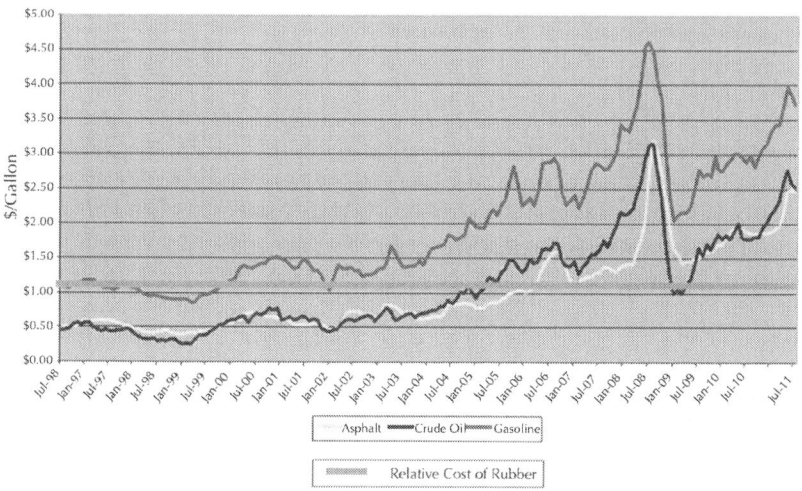

Figure 18: Crude oil, Fuel, Bitumen and rubber cost's trends, adapted from [54].

Lifecycle Economics: As in any economical evaluation, cost effectiveness should be evaluated using Life Cycle Cost Analysis. Again, Hicks and Epps (2003) showed that evaluating different scenarios, in terms of pavement design, maintenance and rehabilitation strategies, the following was concluded:

- High Viscosity RTR-MB (Asphalt-Rubber) is a cost effective alternate for many highway pavement applications mainly due to the ability to reduce thickness when using rubberised asphalt. But also that,

- High Viscosity RTR-MB (Asphalt-Rubber) was not cost effective in all applications.
- When variability is considered in the inputs (cost, expected life, etc.), the Asphalt-Rubber alternates would be the best choice in most of the applications considered.

Other studies conducted at the Arizona State University [47] compared maintenance and user costs trends for the conventional bituminous pavements and Asphalt-Rubber pavements (Fig. 19). Results showed that after 5 years the maintenance and user costs are not much different, after 10 years the maintenance cost begins to substantially be different, as higher maintenance costs will be anticipated for the conventional pavement. This difference for user costs starts at about 15 years. Based on the data analysis presented for the two pavements, an Asphalt-Rubber pavement would be more cost-effective than a conventional pavement with respect to road authorities costs as well as user costs.

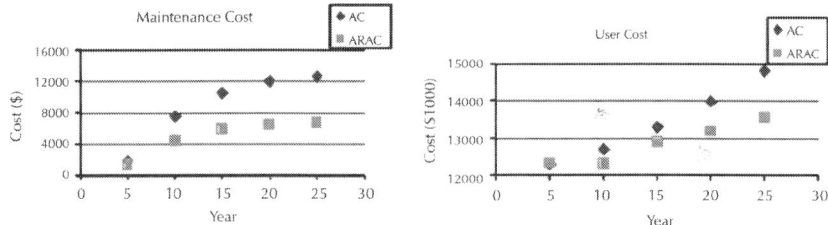

Figure 19: Maintenance cost (left) and User cost comparison between conventional bituminous mixes (AC) and Asphalt-Rubber mixes (ARAC).

Plant Modifications: Another issue that contribute to higher initial cost of the wet process-high viscosity are plant modifications. In case of the production of field blends it is necessary to adapt standard hot mix asphalt plant, for instance with portable units as illustrated in Section 2.3.3. Additionally, conventional paving equipment without modifications is used to place the material. From an investigation made in 2007 in UK by the Waste & Resources Action Programme [18] in order to study a possible adoption of this technology, resulted that the approximate cost for importing/hiring and running the blending plant for 5 to 7 days of rubberised bitumen production is estimated to be between £60,000 and £90,000 (70.000–105.000). This is because none special mixing units described above is currently available in the UK, and therefore requires importing/hiring from either the US or mainland Europe. Moreover, for the installation of a Portable rubberised bitumen mixing plant, an area of not less than 150 m² will be required, depending upon the model and make of the mobile plant [18]. It is worth highlighting that mobilisation and set up of the field-blends binder production equipment cost as much for small jobs as for big ones, but large projects spread mobilisation costs over more asphalt tonnage. The memo suggests that smaller projects may not be cost effective with respect to initial cost. Although the break point for project size may have changed since then, unit costs of small projects (three days paving or less) should be evaluated by LCCA during the design phase [36].

Environmental Issues

High Viscosity RTR-MBs provides the obvious environmental benefit of using waste tyres. Nevertheless, their production and their applications in asphalt presents the following issues:

Hazardous Emissions: Fume emissions have been studied extensively in a number of Asphalt-Rubber projects in USA since 1989. Different studies, performed also by the National Institute for Occupational Safety and Health (NIOSH), Federal Highways Administration (FHWA), [55], determined that use of Asphalt-Rubber does not appear to increase health risks to paving personnel. In these studies the following is reported: "…risks associated with the use of Asphalt-Rubber products were negligible" and also "Emission exposures in Asphalt-Rubber operations did not differ from those of conventional asphalt operations" and also "…the effect of CRM on emissions may be relatively small in comparison to the effects of other variables"[36]. Those variables include the fueling rate of the dryer, mix temperature, asphalt throughput rate and asphalt binder content.

However, some agencies also concluded that the rubber modified mix had an objectionable odour [21]. Moreover, in Colorado while researching and developing specifications for the use of rubberised asphalt for the 2006 paving season, local contractors expressed concerns about using the "wet or dry methods", due to the excessive smoke and smell that would be expelled into the atmosphere during the manufacturing process of this material at their asphalt plants. The contractors were so concerned about losing their state environmental certifications that they indicated they would not use the wet or dry methods without some assurances that their operating permits would not be jeopardised [56]. Nevertheless, several recent studies coupling the RTR-MBs with the Warm mix technology have shown a significant reduction of emission during field operations due to an important decrease of the mixing and compaction temperatures [57] and [21]. Therefore, environmental concerns upon the widespread usage of the wet process still exist, but the development of these technologies is proving that also this issue can be significantly reduced.

Recyclability: Before 1992, in USA Asphalt-Rubber pavements had been performing well and the replacement/recycling of them was not necessary. As some sections of Asphalt-Rubber pavements have met their service life span, recyclability became an issue. Carlson and Zhu [58] report of two jobs occurred in the USA where Asphalt-Rubber mixes were successfully recycled. One example is the recycling job occurred in the City of Los Angeles, California [59], where the initial placement of the Asphalt-Rubber pavement occurred in 1982. In 1994 the pavement was milled and stockpiled at a nearby asphalt plant. The Asphalt-Rubber grindings were added to the virgin rock and oil so that the grindings composed 15% of the final mix. At another location, the grindings were put through a microwave process where nearly 100% of the output was composed of recycled Asphalt-Rubber. This project demonstrated that Asphalt-Rubber can be recycled using either microwave technology or conventional mix design technology. Air sampling during paving and recycling determined that employee exposures to air contaminates were well below the Occupational Safety and Health Administration permissible exposure limits, and in most cases below the detection limits.

WET PROCESS-NO-AGITATION

No-Agitation Recycled Tyre Rubber Modified Bitumen (No-Agitation RTR-MB), technology was first used in Florida and Texas in the mid-1980s and till nowadays it seems to have been used only in some other states of the USA. This is a form of the wet process where CRM is blended with hot bitumen at the refinery or at a bitumen storage and distribution terminal and transported to the asphalt plant/job site for use. In fact the main intuition behind this technology is to take advantage of the engineered properties of the RTR by using the CRM as an alternative modifier to produce storage stable modified bitumen. These binders are often labeled as terminal blends, although nowadays they may also be field-blended at the asphalt plant. Furthermore, this terminology could sound restrictive also because some of the terminal-blended RTR-MBs, with or without addition of polymers, still present the issue of phase separation and need agitation by special augers or paddles. For this reason someone labeled the storage stable terminal blends as terminal blend-hybrid [60]. Therefore a preferred description for this type of binders is "wet process-No-Agitation" blends [36], which clearly indicates the ability of this material of not requiring special equipment to keep the CRM particles evenly dispersed in the modified binder even during the storage process.

Production

In this process, no modifications to the asphalt plants are required because the No-Agitation RTR-MBs are manufactured with specific pressure, temperature, time and agitation requirements similarly to polymer modified bitumen. In the wet process-No-Agitation CRM and bitumen are blended through the continuous blending-reaction systems. No reaction tanks are provided. The bitumen is heated under a controlled environment in a tank to an elevated temperature and high shear stress. The CRM is then introduced into the tank and digested into the bitumen. The characteristic swelling process of the RTR-MB technologies is replaced by the depolymerisation/devulcanisation and optimised dispersion of the rubber into the bitumen by using high processing conditions (temperatures between 200 and 300 °C, shear stresses in the order of few thousands RPM and eventually with pressure higher than 1 atm), resulting in a smooth, homogeneous product. In the past, rubber contents for such blends have generally been 10% by weight of bitumen or total binder, but some products now include even 25% CRM [61].

There are some proprietary and other non-proprietary processes. Some of them also include small percentage of other polymers (e.g.

SBS), or other additives, to produce a combined homogeneous material that exhibits excellent storage stability and compatibility with the finished binder formulation. Such binders are typically modified with CRM particles finer than the 300 μm (No. 50) sieve size that can be digested (broken down and melted in) relatively quickly and/or can be kept dispersed by normal circulation within the storage. During this process, the operator takes samples and runs a solubility test to ensure the rubber is completely digested. Most manufacturers used a high shear process to make sure the CRM is completed digested. The solubility of the finished product is generally above 97.5%. The binder is then delivered to the hot mix plant by truck as a finished product with no additional handling or processing. After mixing with aggregate material is shipped to the job and no special equipment is required for paving, or odour/fume control. Moreover, the RTR mix is compacted like regular hot mix asphalt [62].

Storage Stability

The wet process-No-Agitation aims to produce bitumen-Recycled Tyre Rubber blends which do not occur in eventual phase separation of the CRM from the binder during storage or transportation. The mechanism of phase separation is related to the differences in the properties, mainly physical, of the material constituents in the blends. After production of the blend, the presence of non-dissolved CRM particles in bitumen leads to phase separation, particularly when the blends are stored at high temperatures. As a result, the swollen CRM particles are considered to settle down quickly due to initial higher density than the bitumen phase [63]. On the other hand, migration of the CRM particles to the top of the storage container due to a reduced density after swelling has also been reported in some research studies [64] and [65]. These different mechanisms lead to an unstable condition which results in a rubberised blend with varied properties after the storage.

The improvement of the settling properties of the No-Agitation RTR-MBs is often linked to the careful selection of the blend's components usage of high-curing conditions [66] which ensures a high level of solubilisation of the CRM within the bitumen matrix. Some patented procedure states that as long as the CRM is fully digested into the binder (solubility is above 97%) it is possible to store them without phase separation [62]. However, at this stage the effect of the modification is significantly reduced. In fact, CRM is shown to a number of studies to effectively modifying the bitumen when it is still in the swollen state and not completely devulcanised/depolymerised (Fig. 10c). In order to maintain a convenient level of modification, recent studies have shown that is possible to improve the storage stability of RTR-MBs also by adding substances to operate a chemical stabilisation of the blends [67], by mixing rubberised bitumen blends in presence of a low percentage of polymers [68], with compatibiliser to activate the CRM [69], or with sulphur to improve crosslinking [64]. Another study [70] shows that hot storage stability of rubberised bitumens could be improved by blowing oxygen gases through these blends. In addition, pre-treatment of CRM particles with hydrogen peroxide has also resulted in a more stable RTR-MBs [71] and [72]. Furthermore, Attia and Abdelrahman [68] suggest improving the stability of modified binders also by lowering the storage temperatures. In any case, the stability of No-Agitation RTR-MBs during prolonged storage seems to be similar to other blended PMBs and the man role in improving the settling properties of bitumen-RTR blends is covered by an accurate selection of the processing conditions (Lo Presti et al. 2012).

Properties (Like a PMB)

Although such binders may develop a considerable level of rubber modification, rotational viscosity values rarely approach the minimum threshold of 1500 cPs at 177 °C, that is necessary to significantly increase binder contents above those of conventional asphalt mixes without excessive drain down [36]. Recently in the United States the specifications used for terminal blend (PCCAS, 2008) have utilised the PG grading system, named PG-TR system, similar to the ones used for polymer modified bitumen (Fig. 4.33). In California, PG-TR grades are specifically targeted for use in the same applications for which PG-PMB binders are used, including dense-graded mixes for thick structural sections. The terminal blends now meet the ASTM definition for minimum CRM content [73].

Within the No-Agitation RTR-MBs production, two main properties are of concern: first, the development of performance-related properties and, second, binder compatibility or storage stability. The literature indicates that performance-related properties develop early in the process while compatibility requires few hours to stabilise [74]. Few investigators have conducted studies on the rheological properties of these binders. For example, Thodesen et al. [75] evaluated several binders using the multiple creep recovery tests. Among the binders studied were an Asphalt-Rubber binder and a No-Agitation RTR-MB that uses 10% CRM and 1% SBS polymer. The results showed that the analysed RTR-MB exhibited the least creep and the highest percentage recoveries under various loading and temperature ranges. Another example is given by Attia and Abdelrahman [68], which investigating the possibility of producing high-performance terminal blends, suitable for SuperPave applications, found that by accurately regulating the processing conditions it is possible to balance the development of performance-related properties and storage stability. The same researchers proposed a procedure consisting in blending bitumen with 5% of CRM and 2% of SBS at elevated shear stress (30–50 Hz) and with the processing conditions shown in Fig. 4.38. They found that the final products have performance and binder stability (increasing shearing speed to 50 Hz) comparable to those of patent or proprietary products. Hence, interaction temperature on binder stability was not evaluated. Moreover the authors suggest enhancing the stability of binder by reducing the storage temperature at the plant.

Applications

No-Agitation RTR-MBs can be used with rubber contents as low as 5% and as high as 25%, depending on the application and the project's requirements. These products can be used in all paving and maintenance applications as a replacement of the polymer modified bitumens: [60]:

- Recycled Tyre Rubber Modified Paving Grade Bitumens.
- Standard PG grades – dense graded mixes.
- PG plus grades – high performance mixes.
- Warm mix standard grades – dense and open/gap graded mixes.
- Warm mix polymer modified grades – dense and open/gap graded mixes.
- Hot applied chip seal binders – neat and polymer modified.
- Recycled Tyre Rubber Modified Cutback Bitumens.

- (MC) Medium Cure Graded.
- (SC) Slow Cure Graded.
- Recycled Tyre Rubber Modified Bitumen emulsions.
- Rapid set chip seal.
- Micro surfacing slurry seals.
- Standard slurry seals.
- Cold in-place recycling.
- Cold Mix.
- Recycled Tyre Rubber Modified Seal Coats.
- Recycled Tyre Rubber Modified Surface Sealer.
- Recycled Tyre Rubber Modified Fog Seals.

DISCUSSION AND CONCLUSION

High Viscosity RTR-MBs and No-Agitation RTR-MBs are different products and should not be interchanged (Fig. 20). However, both provide superior cracking performance at reduced thickness, when compared to conventional dense graded hot mix asphalt pavement. Only when there are constructions issues the products are not be expected to perform in a superior manner.

Figure 20: RTR-MB from Wet process (left) and Wet Process-No-Agitation (right), adapted from [73].

In terms of binder properties, the main differences between these products are the viscosity and their storage stability. Viscosity for No-Agitation RTR-MB can range between 500 and 1000 centipoises at 135 °C, much lower than the viscosity for High Viscosity RTR-MB which is in the range of 1500–5000 centipoises at 177.5 °C (Table 8). With regards to the storage stability, No-Agitation RTR-MB born with the idea of obtaining a RTR modified binder comparable to the normal PMBs, therefore on contrary to High Viscosity RTR-MB, it is usually possible to store it as conventional PMBs. Furthermore, a recent study [76] highlights that the storage temperature of the No-Agitation RTR-MB could be significantly decreased (i.e. 15 °C) if compared to the field blended High Viscosity RTR-MB. Considering also the need of the field blends to be stored in agitated tanks, from this point of view the terminal blends leads to significant energy and money savings.

On the other hand, asphalts obtained by using High Viscosity RTR-MBs have more performance history since this process started over in 1960s and they have been used successfully with many applications. With regards to asphalt mixtures, High Viscosity RTR-MB technology is very successful when used with Open-Graded surface courses, where the high air void content of the mix allows an aggregate coating with a much thicker film (36 μm) of high RTR content modified bitumens (about 20%) which leads to an asphalt mix with significantly high binder content (about 7–9%) and with widely proven reduced oxidation, increased durability and increased resistance to reflective cracking. All these benefits are reduced when High Viscosity RTR-MBs is used for Dense-Graded hot mix projects since the dense gradation cannot adequately accommodate the rubber particle size, film thickness is reduced (9 micron) as well as acceptable binder content (about 5%) and the RTR-MBs needs to be produced with much lower rubber content (about 10%). The use of special equipment is not anymore justified by the significant benefits of a thicker coating, therefore in the case of Dense-Graded asphalt mixes the No-Agitation RTR-MBs are the most suitable. On this regard, they are more likely to compete with polymer modified bitumen rather than High Viscosity RTR-MB.

No-Agitation RTR-MBs have been successfully used for a much wider range of products as for instance chips seal applications, open graded and gap graded mixes and emulsions. Basically, No-Agitation RTR-MBs can be used wherever conventional asphalt mixes or asphalt surface treatments are needed. The lower viscosity of No-Agitation RTR-MB implies the usage of less binder per unit area (5–6% binder content) indicating less performance life than if High-Viscosity RTR-MB is used (8–10% binder content). In fact, the ability to inject more binder in the mix translates to better fatigue and reflective cracking performance.

In conclusion, the main advantages of the usage in asphalt mixture of the wet process-No-Agitation in lieu of the wet process – High Viscosity method at the contractor's plant include:

- No need for costly specialised equipment at the asphalt plant.
- No portable plants required for blending of Crumb Rubber with asphalts.
- No additional holding areas for storing the Crumb Rubber product.
- Easiest for the contractor to incorporate into their traditional manufacturing process.
- Mixing, laying and compaction temperatures are comparable to standard mix applications with PMBs.
- Works with all mix designs; does not require any special adjustments to gradation or mix design parameters.
- Being prepared at the refinery, completely eliminates potential problems with heating and blending of Crumb Rubber and asphalt products and
- Eliminates smoke and particulates from entering the atmosphere.
- The binder can be Performance Graded and emulsified
- The few available analyses on the sustainability of the product show that No-Agitation leads to economical saving and environmental benefits.

The aspects that make the wet process-High Viscosity preferable to the wet process-No-Agitation are:

- Evidence of great long term performance, while the long-term performance of existing projects using wet process-No-Agitation are still under evaluation.
- Some have expressed concern that there is no way to verify the amount of CRM used in No-Agitation RTR-MBs.
- Moisture susceptibility of asphalt mixture containing No-Agitation RTR-MB is still not clear.

Conclusions

The author believes that the widespread use of the RTR-MBs technologies within the road pavement industry is advisable. In fact the several benefits provided to the asphalt pavement performance, and to the overall sustainability of the infrastructure, are so evident that it is strongly advised to consider RTR-MBs technologies as a first option to the binders currently used in road pavements. Companies, road authorities, etc. have to evaluate if it is convenient to use the High Viscosity wet process technology, which proved widely to provide several benefits, in particular it allows highway designers to reduce pavement layer thickness due to the proven properties of rubberised bitumen, but presents some challenges as: the need for suitable blending and mixing equipment, the cost of such equipment and the degree of difficulty in preparing asphalt mix design. The other option is to choose the wet process-No-Agitation technology which solves several issues but leads to asphalts pavements with, so far, uncertain long term performance. In both cases the implementation of these technologies still presents issues as: the lack of availability of suitable RTR processing facilities in the vicinity and the cost of such facilities, the lack of rubberised bitumen binder and mix standards, the uncertainty on the environmental performance of these products and the lack of trained personnel particularly for accurate laboratory analyses, of raw materials and final blends, which are necessary to optimise the modification process and the performance of the final product. Education and training are indeed another key aspect for a successful application of RTR-MBs in road asphalt mixtures also because new procedures are being developed to perform an enhanced laboratory design of these technologies [77] as well as reducing limitations to their applicability in the field. Examples are given by the increasing number of rubberised asphalt applications with additives able to reduce paving temperatures (i.e. warm mix technologies) [21] and [57], but also from the invention of a new form of product delivery as the case of the rubberised pellets [78]. Another important aspect to clarify is the cost of this technology: it is true that initial costs are still an issue but the rapid cost increase of bitumen and results of lifecycle cost analyses indicate that RTR-MBs are in many cases an economically convenient option. At last, there is evidence that a closer involvement of local and national governments with policies supporting the RTR industries and the major investments on training and research could definitely help to decrease the costs of implementing RTR-MB technologies as well as solving the several issues indicated within this manuscript. A proof of this statement is given by the state of California (USA) where since

2005 a government mandate [79] supports the increasing adoption of these technologies in asphalt mixture and from the January 2013 the usage of rubberised asphalt pavements needs to reach at level of at least 35% of the total weight of asphalt paving materials produced in the country. The main reason of this mandate is the prediction of substantial savings in the long term.

REFERENCES

1.   ETRMA statistics. <http://www.etrma.org/pdf/20101220%20 Brochure%20ELT_2010_final%20versio.pdf>; 2012.

2.   ETRMA ELTs. End-of-Life Tyres. <http://www.etrma.org/uploads/ Modules/Documentsmanager/brochure-elt-2011-final.pdf>; January 2011.

3.   Rahman MM. Characterisation of dry process crumb rubber modified asphalt mixtures. Thesis submitted to the University of Nottingham for the degree of Doctor of Philosophy. s.l.: University of Nottingham, School of Civil Engineering; 2004.

4.   N. Miskolczi, R. Nagy, L. Bartha, P. Halmos, B. Fazekas, Application of energy dispersive X-ray fluorescence spectrometry as multielemental analysis to determine the elemental composition of crumb rubber samples, Microchem J, 88 (1) (2008), pp. 14–20

5.   Diffusion kinetics of bitumen into waste tyre rubber. Artmandi, I and Khalid, H A. Savannah: s.n., 2006. Journal of the Association of Asphalt Paving Technologists. Proceedings of the Technical Sessions. vol. 75. Georgia, USA; 2006. p. 133–64.

6.   Peralta E. Study of the interaction between bitumen and rubber – PhD thesis. <http://repositorium.sdum.uminho.pt>; 2009 [Cited: 08.03.13] <http://repositorium.sdum.uminho.pt/ bitstream/1822/10557/1/Tese_Mestrado_ER_Joana.pdf>.

7.   Checkthatcar.com. <http://www.checkthatcar.com>; 2013.

8.   Rimondi G. <http://www.acrplus.org/upload/documents/ events/2009_Turin_Rimondi.pdf>; 2009.

9.   Walker D. Understanding how tires are used in asphalt. <http:// www.rubberform.com/news.php?id=108>; 2013.

10.   Oliver JWH. Modification of paving asphalts by digestion with

scrap rubber. [ed.] Transportation Research Board. Transportation Research Record 821; 1981.

11. F.L. Roberts, P.S. Kandhal, E.R. Brown, R.L. Dunning, Investigation and evaluation of ground tyre rubber in hot-mix asphalt, NCAT Auburn University, Alabama, USA (1989)

12. Memon N. Characterisation of conventional and chemically dispersed crumb rubber modified bitumen and mixtures. University of Nottingham. Nottingham, UK: s.n. PhD thesis; 2011.

13. Heitzman M. Design and construction of asphalt paving materials with Crumb Rubber Modifier. Transportation Research Record 1339; 1992.

14. Epps JA. Uses of recycled rubber tyres in highways. Washington, DC: Synthesis of Highway Practice No.198, TRB National Research Council. NCHRP Report; 1994.

15. Kuennen T. Asphalt rubber makes a quiet comeback. Better Roads Magazine; May 2004.

16. Terrel RL, Walter JL. Modified asphalt materials – the European Experience. In: s.l.: AAPT proceedings. vol. 55. 1986. p. 482–518.

17. Bitumen–rubber: lessons learned in South Africa. Visser AT, Verhaege B. Portugal: ISBN: 972-95240-9-2, 2000. Asphalt Rubber; 2000.

18. Widyatmoko I, Elliot R. A review of the use of crumb rubber modified asphalt worldwide. UK : Waste & Resources Action Programme (WRAP); 2007.

19. Souza R. Experiences with use of reclaimed rubber in asphalt within Europe. Birmingham: s.n. Rubber in Roads; 2005.

20. Mavridou S, Oikonomou N, Kalofotias A. Worldwide survey on best (and worse) practices concerning rubberised asphalt mixtures implementation (number of different cases, extent of application. Thessaloniki: EU-LIFE+ Environment Policy and Governance. ROADTIRE, D2.1.1; 2010.

21. Hicks G, Cheng D, Teesdale T, Assessment of Warm Mix technologies for use with Asphalt Rubber paving application. Tech-Report-103TM; 2010.

22. Antunes M, Baptista F, Eusébio MI, Costa MS, Valverde Miranda

C. Characterisation of asphalt rubber mixtures for pavement rehabilitation projects in Portugal. Asphatl Rubber 2000; 2000.

23. Gallego J, Del Val MA, Tomas R. Spanish experience with asphalt pavements modified with tire rubber. Asphalt Rubber 2000; 2000.

24. Santagata FA, Canestrari F, Pasquini E, Mechanical characterisation of asphalt rubber – wet process. Palermo: s.n. In: 4th International SIIV Congress; 2007.

25. Dasek O, Kudrna J, Kachtík J, Spies K. Asphalt rubber in Czech Republic. Munich: s.n. Asphatl rubber 2012; 2012.

26. Nordgren T, Tykesson A, dense graded asphalt rubber in cold climate conditions. Munich: s.n. Asphalt Rubber 2012; 2012.

27. Pinto A, Sousa J, The first brazilian experience with in situ field blend rubber asphalt. Munich: s.n. Asphalt Rubber 2012; 2012.

28. Nourelhuda M, Ali G. Asphalt-rubber pavement construction and performance: the sudan experience. Munich: s.n. Asphalt Rubber 2012; 2012.

29. Bahia H, Davis R, Effect of Crumb Rubber Modifiers (CRMs) on performance related properties of asphalt binders. AAPT 1994; 1994.

30. Cheovits JG, Dunning RL, Morris GR. Characteristics of asphalt-rubber by the slide plate microviscometer. Association of Asphalt Paving Technologists. vol. 51. 1982. p. 240–61.

31. Review of Utilization of Waste Tires in Asphalt. Shatanawi K, Thoedsen C. s.n. In: Globhal Plasti Environmental Conference. Orland, FL, USA; 2008.

32. M.A. Abdelrahman, S.H. Carpenter, The mechanism of the interaction of asphalt cement with crumb rubber modifier (CRM), Transport Res Rec, 1661 (1999), pp. 106–113

33. D. Lo Presti, G. Airey, P. Partal, Manufacturing terminal and field bitumen-tyre rubber blends: the importance of processing conditions, Proc-Soc Behav Sci, 53 (2012), pp. 485–494

34. DGRoadMachinery.com. www.dgroadmachinery.com. <http://www.dgroadmachinery.com/3-3-asphalt-rubber-blending-plant.html>; 2013.

35. WRIGHTasphalt. Terminal blendedterminal blendedtire rubber modified tire asphalt cementasphalt cement. <http://nwpma-online.org/resources/08Spring_Trmss.pdf>.

36. Caltrans. Asphalt Rubber Usage Guide. s.l.: State of California Department of Transportation, Materials Engineering and Testing Services; 2006.

37. ASTM D6114/D6114M. s.l.: American Society for Testing and Materials; 2009. Standard Specification for Asphalt-Rubber Binder.

38. AsAc. The use of Modified Bituminous Binders in Road Construction. Technical Guideline of the South African Asphalt Acadamey TG1; 2007.

39. Potgieter CJ, Bitumen Rubber Asphalt in South Africa Conventional Techniques. Vienna: s.n. In: 3rd Eurasphalt & Eurobitume Congress; 2004.

40. SABITA. manual 19. Guidelines for the design, manufacture and construction of bitumen rubber asphalt wearing courses; 2009.

41. Austroads. Specification framework for polymer modified binders and multigrade bitumens". Austroads Technical Report AP-T41/06; 2006.

42. OGFC Meets CRM. Where the Rubber meets the Rubber. 12 Years of Durable Success. Way, G.B. Vilamoura: s.n. In: Asphalt Rubber 2000 Conference; 2000.

43. Santagata FA, Antunes I, Canestrari F, Pasquini E. Asphalt Rubber: Primeiros Resultados em Itália. Estoril: s.n. Estrada 2008, V Congresso Rodoviário Português; 2008.

44. Sousa JB, Experiences with use of reclaimed rubber in asphalt within Europe. Birmingham: s.n. Rubber in Roads; 2005.

45. Bertollo SM, Mechanical properties of asphalt mixtures using recycled tyre rubber produced in Brazil – a laboratory evaluation. Washington: s.n. In: 83rd TRB Annual Meeting; 2004.

46. Kaloush KE, Witczak MW, Sotil AC, Way GB, Laboratory evaluation of asphalt rubber mixtures using the dynamic modulus ($E*$) Test. Washington: s.n. In: 82nd TRB Annual Meeting; 2003.

47. Jung J, Kaloush KE, Way GB, Life cycle cost analysis: conventional versus asphalt rubber pavements. s.l.: Rubber Pavement Association., 2002; 2003.

48. Noise Reducing Asphalt Pavements: a Canadian Case Study. Leung F et al. s.n. In: 10th International conference on asphalt pavements. Quebec City; 2006.

49. The Successful World Wide Use of Asphalt Rubber. Antunes I et al. Cosenza: s.n. In: 16th Convegno Nazionale SIIV; 2006.

50. Pasquini E. Advanced characterisation of innovative environmentally friendly bituminous mixtures. s.l.: Università Politecnica delle Marche. PhD thesis; 2009.

51. Antunes I, Asphalt rubber: Il Bitume Modificato con Polverino di Gomma di Pneumatico Riciclata. L'Aquila: s.n. Varirei – V International Congress of Valorisation and Recycling of Industrial Waste; 2005.

52. Advances in technology of asphalt paving materials containing used tyre rubber. Takallou HB and Sainton A. Transportation Research Reccords. vol. 1339. 1992. p. 23–29.

53. Hicks R, Epps JA. Life cycle costs analysis of asphalt rubber paving materials. Tempe, AZ: The Rubber Pavements Association; 2003.

54. Carlson D. Developing trends in rubberized asphalt. US EPA SMM Web Academy Webinar Series, www.epa.gov. <http://www.epa.gov/smm/web-academy/2013/pdfs/smm_webinar_022113_carlson.pdf>; 2013.

55. Crockford WW, Recycling crumb rubber modified asphalt pavements. s.l.: Texas Transportation Institute. Report FHWA/TX-95/1333-1F; 1995.

56. Khattak S, Syme B. Terminal blend tyre rubber asphalt technical information. s.l.: City of Colorado Springs and Nolte Associates, Inc.; 2010.

57. Jones D, Wu R, Barros C, Peterson J. Research findings on the use of rubberised warm-mix asphalt in California. Asphalt-Rubber 2012; 2012.

58. Carlson D, Zhu H, An anchor to crumb rubber markets. <http://www.rubberpavements.org/Library_Information/4_5_Asphalt-Rubber_an_Anchor_to_Crumb_Rubber_Markets.pdf>; 1999.

59. Youssef Z, Hovasapian PK. Olympic boulevard asphalt rubber recycling project. s.l.: City of Los Angeles Department of Public Works; 1995.

60. Quire D. Tire rubber modified asphalt, asphalt pavement conference presentations. <http://asphaltroads.org/images/documents/tire-rubber-modified-asphalt.pdf>; March 2012.

61. Raetexindustries.com, Terminal_Blend_vs_AR.pdf. <http://raetexindustries.com/blog/wp content/uploads/2011/10/Terminal_Blend_vs_AR.pdf>.

62. Asphalt Institute online magazine, Terminal Blended Rubberised Asphalt Goes Mainstream – Now PG Graded. <http://www.asphaltmagazine.com/singlenews.asp?item_ID=1607&comm=0&list_code_int=mag01-int>; 2008.

63. F.J. Navarro, P. Partal, F.J. Martinez-Boza, F. Gallegos, Thermo-rheological behaviour and storage stability of ground tyre rubber-modified bitumens, Fuel, 83 (14–15 October 2004)

64. N.F. Ghaly, Effect of sulfur on the storage stability of tire rubber modified asphalt, World J Chem, 3 (2) (2008), pp. 42–50

65. A. Perez-Lepe, F.J. Martinez-Boza, F. Gallegos, High temperature stability of different plymer-modified bitumens: a rheological evaluation C, J Appl Polym Sci (2007)

66. Glover CJ, Davison RR, Bullin JA, Estakhri CK, Williamson SA, Billiter TC, Chipps JF, Chun JS, Juristyarini P, Leicht SE, Wattanachai P. A comprehensive laboratory and field study of high-cure crumb-rubber modified asphalt 21 materials. USA: Texas Transportation Institute; 2000. Report 1460-1.

67. Biro S, Bartha L, Deak G, Geiger A, Pannonn E. Chemically stabilized asphalt rubber compositions and a mechanochemical method for preparing the same. HU226481, WO/2007/068990 – 2007, EP1960472 – 2008; 2008.

68. M. Attia, M.A. Abdelrahman, Enhancing the performance of crumb rubber-modified binders through varying 1 the interaction conditions. M. 6, Int J Pavement Eng, 10 (2009), pp. 423–434

69. G. Cheng, B. Shen, J. Zhang, A study on the performance and storage stability of crumb rubber-modified 11 asphalts, Petrol Sci Technol (2010)

70. William GE, Polymer enhanced asphalt. WO patent, 97455488; 1997.

71. Maldonado PJ, Phung TK, Process for preparing polymers-bitumen compositions. US patent, 4145322; 1979.

72. Planch JP. Method for the preparation of bitumen polymer compositions. WO Patent, 9002776; 1990.

73. Shatnawi S, Comparisons of runnerized asphalt binders: asphalt-rubber and terminal blend. Munich: s.n. In: Asphalt Rubber proceedings 2012; 2012.

74. M.A. Abdelrahman, S.H. Carpenter, The mechanism of the interaction of asphalt cement with crumb rubber modifier (CRM). USA : s.n. Trans Res Rec, 1661 (1999), pp. 106–113

75. Thodesen C, Biro S, Kay J. Evaluation of current modified asphalt binders using the multiple stress creep recovery test. s.n. In: Proceedings of AR2009 conference. Nanjing, China; 2009.

76. Wu C, Liu K, Tang J, Li A. Research on the terminal blend rubberized asphalt with high-volume of rubber crumbs and its gap graded mixture. Munich: s.n. Asphalt rubber 2012; 2012.

77. B. Celauro, C. Celauro, D. Lo Presti, A. Bevilacqua, Definition of a laboratory optimization protocol for road bitumen improved with recycled tire rubber. s.l, Constr Build Mater, 37 (2012), pp. 562–572

78. Amirkhanian S, Kelly S. Development of polymerized asphalt rubber pelleted binder for HMA mixtures. Munich: s.n. In: Asphalt rubber proceedings 2012; 2012.

79. Leginfo.ca.gov. <http://www.leginfo.ca.gov/pub/05-06/bill/asm/ab_0301-0350/ab_338_cfa_20050427_134930_asm_comm.html>; 2013.

Citations

CHAPTER 1

Simone Pascucci, Cristiana Bassani, Angelo Palombo, Maurizio Poscolieri and Rosa Cavalli, Road Asphalt Pavements Analyzed by Airborne Thermal Remote Sensing: Preliminary Results of the Venice Highway, ISSN 1424-8220.

CHAPTER 2

Fidelis O. OKAFOR, Performance of Recycled Asphalt Pavement as Coarse Aggregate in Concrete, ISSN 1583-1078.

CHAPTER 3

Kiplagat Chelelgo, Effects of Diurnal Temperature Dynamics on Curing of Cold-Emulsion Reclaimed Asphalt Pavements, ISSN 1819-6608.

CHAPTER 4

Ankit Gupta, Abhinav Kumar, comparative Structural Analysis of Flexible Pavements Using Finite Element Method, doi: 10.2478/ijpeat-2013-0005

CHAPTER 5

O. Reyes-Ortiz, E. Berardinelli, A.E. Alvarez, J.S. Carvajal-Muñoz and L.G. Fuentes, Evaluation of hot mix asphalt mixtures with replacement of aggregates by reclaimed asphalt pavement (rap) material, doi:10.1016/j.sbspro.2012.09.889.

CHAPTER 6

Manoj Shukla / Dr. Devesh Tiwari / K. Sitaramanjaneyulu, Performance Characteristics of Fiber Modified Asphalt Concrete Mixes, doi: 10.2478/ijpeat-2013-0007.

CHAPTER 7

Marco Pasetto / Giovanni Giacomello, Experimental Analysis of Waterproofing Polymeric Pavements for Concrete Bridge Decks, doi: 10.2478/ijpeat-2013-0008.

CHAPTER 8

Zhang Junwei, Li Jinping, and Quan Xiaojuan, Thermal Stability Analysis under Embankment with Asphalt Pavement and Cement Pavement in

Permafrost Regions, The Scientific World Journal, vol. 2013, Article ID 549623, 12 pages, 2013. doi:10.1155/2013/549623.

CHAPTER 9

Altan Cetin, Effects of Crumb Rubber Size and Concentration on Performance of Porous Asphalt Mixtures, International Journal of Polymer Science, vol. 2013, Article ID 789612, 10 pages, 2013. doi:10.1155/2013/789612.

CHAPTER 10

Mehmet Cetin (2013). Landscape Engineering, Protecting Soil, and Runoff Storm Water, Advances in Landscape Architecture, Dr. Murat Ozyavuz (Ed.), ISBN: 978-953-51-1167-2, InTech, DOI: 10.5772/55812.

CHAPTER 11

V. Mouillet, D. Séjourné, V. Delmotte, H.-J. Ritter, and D. Lesueur, Method of Quantification of Hydrated Lime in Asphalt Mixtures, doi:10.1016/j.conbuildmat.2014.06.063.

CHAPTER 12

Davide Lo Presti, Recycled Tyre Rubber Modified Bitumens for Road Asphalt Mixtures: A Literature Review, doi:10.1016/j.conbuildmat.2013.09.007

Index